Lecture Notes in Economics and Mathematical Systems
Managing Editors: M. Beckmann and H. P. Künzi

142

John S. Lane

On Optimal Population Paths

Springer-Verlag
Berlin · Heidelberg · New York 1977

Author
John S. Lane
London School of Economics
and Political Science
Aldwych
London WC2/England

Library of Congress Cataloging in Publication Data

Lane, John S
On optimal population paths.

(Lecture notes in economics and mathematical systems ; 142)
Bibliography: p.
1. Population. 2. Saving and investment.
3. Utilitarianism. I. Title. II. Series.
HB885.L25 301.32 77-704

AMS Subject Classifications (1970): 90A15

ISBN-13: 978-3-540-08070-1 e-ISBN-13: 978-3-642-95291-3
DOI: 10.1007/978-3-642-95291-3

PREFACE

The overall purpose of this monograph is to integrate and critically evaluate the existing literature in the area of optimal joint savings population programs. The existing diverse presentations are all seen to be discussions within a unified framework.

The central problem is to compare the desirability of alternative inter-temporal sequences of total savings and population sizes. Of critical importance is whether one regards persons as the fundamental moral entities or whether one takes Sidgwick's viewpoint that something good being the result of one's action is the basic reason for doing anything.

The latter viewpoint is consistent with defining a complete social preference ordering over these alternative sequences. Since part of one's interest is to evaluate the consequences of various ethical beliefs a comparative study of several such orderings is presented; in particular the Mill-Wolfe average utilitarian, and Sidgwick-Meade classical utilitarian, formulations.

A possible problem with the social preference ordering approach is that the ordering may indicate the desirability of increasing the population size, if this increases the total amount of good, even though people may receive less than the welfare subsistence level of consumption. However, there are other ways of evaluating actions and, if persons are the fundamental moral entities, then perhaps these actions should be evaluated by their implications for the rights of individuals i.e. people who are currently alive, people who one can predict will exist in the future (e.g. as given by an exogenous population profile), but not 'potential souls' who only might exist in the future.

In this event, then, the social ordering approach is dubious simply because the domain of aggregation is no longer well-defined. An entirely different framework is required for evaluating alternative inter-temporal savings population programs. An alternative approach which utilizes the Nash non-cooperative solution concept, so providing only a partial ordering over these sequences, is therefore presented.

'First best' solutions in which population size is directly controllable are compared with their 'second best' counterpart in which population growth is endogenous and therefore serves as a constraint on the choice of an optimal savings program. The analysis proceeds under alternative assumptions about the form of the population growth rate function and the implications for an explicit Government population program are considered. This analysis is further extended, albeit tentatively, to incorporate an endogenous dependency ratio.

This monograph was written as part of a project on "Dynamic World Systems Models," at the University of Wisconsin, under the financial support of the Department of Health, Education and Welfare. My thanks are extended to Richard H. Day for enabling me to work on this project; also to Partha Dasgupta and Tjalling Koopmans, on whose work I have drawn extensively, for the many discussions we have had on the subject and also for their encouragement.

TABLE OF CONTENTS

1. INTRODUCTION.

In the last seventy years there has been a rapid acceleration in the rate of growth of the world population. Between the years 1750 to 1900 the population growth rate for the world as a whole was 10% per decade. The United Nations projects a rate of 17% per decade for the next 30 years which will lead to the population doubling in only 40 years. Therefore studying both the causes and possible consequences of high population growth rates is crucial. Through its effect on the process of capital accumulation rapid population growth obstructs efforts to raise the standard of living in developing countries. It alters the age-structure of the population so that the proportion of the population not included in the work force increases. It reduces population-resource ratios and so has implications with respect to the process of production, the distribution of income and environmental quality. Furthermore, to the extent that population growth induces urbanisation it concentrates existing poverty and frustration and thereby exacerbates social unrest and political instability.

The consequences of variable population growth rates should be properly evaluated in the context of optimal economic policies derived from alternative social welfare functions. This is our main concern and in sections 2 to 5 we will introduce control-theoretic models embodying controllable and/or endogenous population growth rates. These models emphasise the importance of the interdependence between population change and the extensive margin of development, the choice of social welfare function and the existence of resources in fixed supply.

High population growth rates reduce the proportion of output that can be allocated to investment in capital goods because a larger slice is required to meet current consumption needs. Also, of the output that is allocated to investment a larger proportion must be used extensively; that is, to equip the new members of the labor force with the same amount of capital per head as existing members. The net result is that a smaller proportion of output is available for capital accumulation and so also for raising productivity levels.

The choice of an optimal savings-population program will be sensitive to the form of the social welfare function. Since part of our aim is to evaluate the consequences of various ethical beliefs a critical and comparative study of several such functions will be made. Those that can be described as "average utilitarian" are considered in section 2 and those that can be described as "classical utilitarian" in sections 3 and 4.

Whilst it is not realistic to assume that population growth rates or the absolute size of the population are instantaneously and costlessly controllable, the consequences of such an assumption have much normative significance. This 'first best' approach to the problem is considered in section 3 and a 'second best' approach (so named because population is assumed to grow endogenously and therefore serves as a constraint on the choice of an optimal savings program) in section 4. Certain modifications of this basic model are also considered ; in particular, in Section 5, exhaustible resources are included.

It should be emphasized that the analysis presented in section 2 to 5 inclusive relates to an aggregate model of a centrally planned economy and as such does not reflect the divergence between private and social costs and benefits that can occur in a mixed or laissez-faire economy.

This divergence creates a difference between how many children parents wish to have and how many they should have. An alternative framework of analysis is presented in section 6 which, in a particular sense, has something to say on this matter. More specifically, however, the purpose in section 6 is to present a case against the use of any type of social welfare function in determining an 'optimal' savings-population program. The relationship between the analysis of sections 2 to 5 inclusive and that of section 6 is also considered.

In summary, the overall purpose of this paper is to integrate and critically evaluate the existing literature in the area of optimal joint savings population programs. In fact, the existing diverse presentations in this area will all be seen to be discussions within a unified framework.

2. AVERAGE UTILITARIANISM.

Utilitarianism provides an apparatus which gives meaning to the concept of an optimal population size or growth rate. For it is natural to define a social welfare function in terms of the utilities of the individuals that constitute society. And then, not only does the concept of an optimum have real significance, but within this framework the merits of alternative savings-population programs can be considered.

The modern conception of an optimum population is generally attributed to Cannan and Wicksell [5,13]. Partly because it is the Cannan-Wicksell formulation which has dominated thinking on this issue for the last century, and partly because it will help clarify certain conceptual issues, it is of importance to develop our analysis from this base.

Consider an economy with a single, homogeneous, non-deteriorating commodity. The stock of this commodity is considered fixed at K_o. Total population L is identified, for simplicity, with the working force. Efficient production Q of the commodity is represented by the functional relationship

$$Q = F(K_o, L)$$

It is reasonable to suppose that output increases when the amount of labor employed increases. It is also supposed that the rate at which this happens first increases due to the benefits of scale but then decreases at higher levels of employment due to congestion, i.e.

$$F_L > 0 \text{ for all } L \geq 0$$

and there exists $\bar{L}$ such that,

$$L < \bar{L} \text{ implies } F_{LL} > 0$$

and vice-versa. Such a technology is illustrated in Figure 1 below. Also illustrated is the implied average product curve ($q \equiv F(K_o, L)/L$) and the implied marginal product curve (F_L).

The Cannan-Wicksell formulation defines the optimum population as being the population size which maximises the average product of labor; one searches for the peak of the bell-shaped average product curve. The optimal population L_c will then have the property:

$$q = F_L$$

or, the average product of labor is equal to its marginal product.

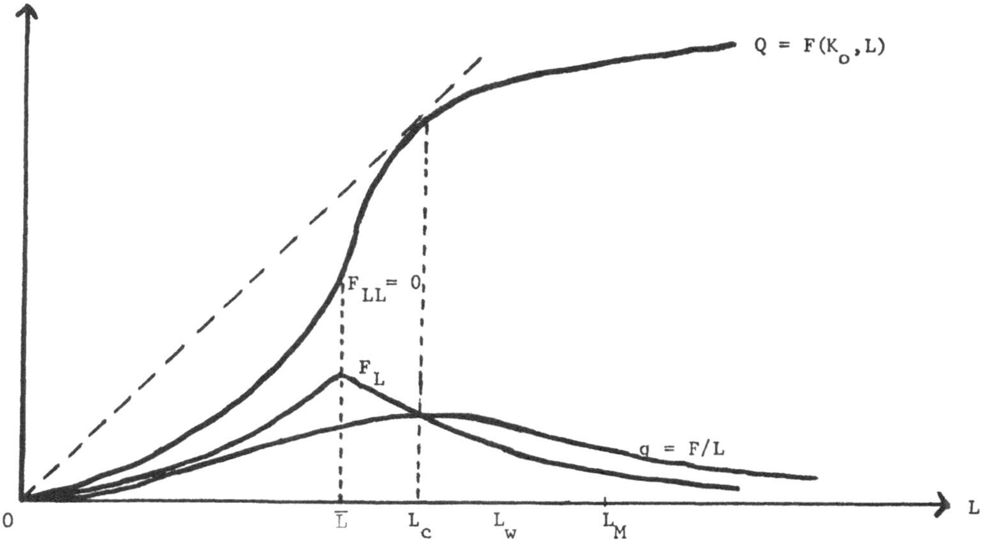

FIGURE 1.

Presumably it is well-being rather than per-capita output which is the proper aim of policy. Furthermore one's concern is to find a systematic measure of how desirable a population and savings policy is in so far as it has various economic features. The interpretation of an optimum is in terms of maximising well-being to the extent that it is a function of the economically accessible variables. Therefore the proposal to regard L_c as an optimum is defensible only if people's welfare or utility is itself thought of as being maximised with per-capita output, i.e.

$$W = u(q) \qquad u'(q) > 0$$

where W is the social welfare function and $u(\cdot)$ the utility function of a representative member of society.

It is true that per-capita output is an obvious candidate for the argument of the social welfare function. So much so that even quite recently Sauvy, in his monumental work [37], restricted his study to this formulation. However this formulation is questionable and it was Wolfe [44,45], writing thirty years earlier, who recognized that the benefits of producing commodities comes only in so far as we consume them, i.e. $W = u(c)$ where $c \equiv C/L$. This formulation is essentially the average utilitarianism of John Stuart Mill. Now to choose L to maximise this social welfare function requires one to specify how much will be saved, S_o say. As:

$$c = q - S_o/L$$

the optimal population size L_w must satisfy

$$q = F_L + S_o/L$$

so $c = F_L$, i.e. at the Wolfe optimum per-capita consumption is set equal to the marginal product of labor. Clearly if $S_o > 0$ then $L_w > L_c$ and it is trivial that $S_o = 0$ implies $L_w = L_c$.

The choice of today's population size depends on the level of savings S_o; so also does the choice of future population sizes. Therefore the question immediately arises as to how S_o should be determined. At this point one notices in a precise way that the problems of optimum population and optimum savings are inter-dependent.

Now there are good reasons for believing that many of the earlier writers were contemplating the optimum stationary economy. Given this interpretation it is implicitly being postulated that both population and per-capita consumption are assumed to remain constant over time. The purpose of the analysis is, then, to determine the optimum constant population and per-capita consumption profiles, i.e. to determine the optimal stationary economy. Then S_o must be set at the level required to compensate for capital depreciation. In the models considered above there is no depreciation so $S_o = 0$ and $L_c = L_w$; the Cannan-Wicksell and Wolfe solutions are the same.

It is natural to ask why, in locating the optimum, one's choice should be restricted to stationary economies. For it should not be a starting assumption of the analysis that a stationary economy is judged optimal but, rather, it should be implied by ethical and moral considerations. Such considerations may be formalized by the choice of an inter-temporal social welfare function. Unfortunately it is not immediate what the intertemporal analogue of the Mill-Wolfe average utilitarianism is.

It has been argued that the economic welfare of a community ought to be reflected by the intertemporal discounted sum of welfares per head, i.e.

(A) $$W_A = \int_0^{\infty} e^{-\rho t}\, u(c_t)dt$$

where $\rho \geq 0$ is the subjective rate of time reference. In order to appreciate the a priori ethical foundations of such a welfare function consider, in particular, the following problem [1]. A country consists of identical individuals (and so they have identical utility functions) located on two islands with populations L_1 and L_2. This country has no means of production but is endowed with a given stock C of some commodity. Now suppose that the government must distribute this stock C amongst the population in such a way as to maximise social welfare,

$$W = u(c_1) + u(c_2)$$

subject to the feasibility constraint,

$$C = L_1c_1 + L_2c_2 \ .$$

Assuming diminishing marginal utility, or $u''(c) < 0$, the solution must satisfy:

$$u'(c_1)/u'(c_2) = L_1/L_2$$

Therefore, if $L_1 > L_2$ then $c_1 < c_2$; per-capita consumption is smaller for someone who happens to locate on the more populated island. The inter-temporal analogue of this result is that the social welfare function (A) will discriminate against future generations if they happen to be large and vice-versa. Discrimination is based solely on size. One can legitimately question the blatant asymmetry in which the claims of different people are being considered in these investigations. Indeed, if it is an inter-temporal version of the average utilitarian principle that one wants, then symmetry of treatment over time dictates the use of a welfare function that is the inter-temporal sum of the total discounted utilities of each generation, divided by the sum of each generation's population size, i.e.

(B) $$W_B = \int_0^T e^{-\rho t}[L_t u(c_t) / \int_0^T L_\tau d\tau]dt. \qquad \rho \geq 0$$

where T is finite or infinite and W_B is invariant under linear increasing transformations of the utility index if $\rho = 0$. Whilst it may be objected that there will be systematic discrimination against future generations if there is positive discounting ($\rho > 0$) of their utilities one should observe that this is not an argument about the relative merits of welfare functions A and B. More fundamentally this systematic discrimination against future generations stems from the a priori ethical belief that future generations are less important than the present generation as seen from the present. In contrast the arbitrary discrimination provided by welfare function A against future generations if they happen to be large (which in any case would be a result of decisions made by the present generation) stems from no such belief; it is simply an unplanned-for side effect.

Modifying one's valuations in the light of their consequences is often a legitimate procedure; part of our purpose, here, is to evaluate the consequences of various ethical beliefs as represented by alternative welfare functions. One would wish for a more robust welfare function which is not only a priori ethically defensible but which also leads to results which accord with our moral intuition in the context of alternative idealized economies.

To illustrate consider the problem of maximizing the social welfare function A with $\rho \geq 0$ and over an arbitrary finite horizon T say. Suppose that today's population L(0) and capital stock K(0) are given but that all population growth rates n_t, including n_o, are controllable (instantaneously and without cost) subject only to lying between prescribed upper and lower limits so,

$$n_* \leq n_t \leq n^*.$$

Also assume that the technology exhibits constant returns to scale together with positive and diminishing marginal products so,

$$q \equiv Q/L = F(K,L)/L = F(K,1) \equiv f(k)$$

where $k \equiv K/L$, $f'(k) > 0$ and $f''(k) < 0$ and suppose that $f(0) = 0$. The capital accumulation constraint may be written, where 'I' denotes gross investment and there is no depreciation,

$$\dot{K}/L = I/L = [F(K,L) - C]/L = f(k) - c$$

so,

(1) $$\dot{k} = f(k) - nk - c$$

where the dot denotes differentiation with respect to time and the implicit variable 't' has been subsumed. Additional feasibility constraints are:

(2) $c \geq 0 \quad k \geq 0$ for all time

and it is assumed that,

$$u'(0) = \infty \qquad u'(\infty) = 0 \qquad u'(c) > 0 \qquad u''(c) < 0$$

which is sufficient to guarantee that c_t always takes an interior solution. This is a simplifying assumption which can be modified without difficulty.

Mathematical techniques for solving such a problem have been developed by Pontryagin et al [32] and illustrated in an economic context by K.J. Arrow [1] amongst others. Application of these techniques requires us to form the current value Hamiltonian 'H' for this problem. By definition:

$$H = u(c) + p[f(k) - nk - c] + \mu_1[n - n_*] + \mu_2[n^* - n]$$

where p is the Pontryagin auxiliary variable associated with the state of the system, namely 'k'. The economic interpretation of p_t is the increase in the value of the social welfare function resulting from a small increase in the capital-labor ratio 'k' at time t, i.e. it plays the same role as a scarcity or shadow price. The variables μ_1 and μ_2 are the Lagrange multipliers associated with the constraints. It is then the case, in particular, that at any moment of time the controls (c and n) must be chosen to maximise the current value Hamiltonian so;

$$\partial H/\partial n = -pk + \mu_1 - \mu_2 = 0$$

$$\partial H/\partial c = u'(c) - p = 0 \quad \text{as } u'(0) = \infty$$

Such controls are indeed maximising because $\partial^2 H/\partial n^2 = 0 = \partial^2 H/\partial n \partial c$ and $\partial^2 H/\partial c^2 = u''(c) < 0$ by assumption. Also it is required that μ_1 and μ_2 be non-negative and $\mu_1[n-n_*] = \mu_2[n^*-n] = 0$. Therefore there are these cases to consider;

(i) $n = n^*$, $\mu_1 = 0$, $\mu_2 \geq 0$

(ii) $n_* < n < n^*$, $\mu_1 = \mu_2 = 0$

(iii) $n = n_*$, $\mu_1 \geq 0$, $\mu_2 = 0$.

If (i) holds then $pk = -\mu_2 \leq 0$ but $p = u'(c) > 0$ so $k \leq 0$. But if ever $k_t = 0$ then it must become negative i.e. this program becomes infeasible in finite time. The same is true if (ii) holds.

This shows that an optimal program (which can easily be shown to exist) must have the property that the population growth rate be at its lower limit for all time. It is reasonable to suppose that $n_* < 0$ so this result states that population must be run down as fast as possible in order to offer higher per-capita consumptions to the lucky few that survive such a program. Furthermore if there is no lower limit on n, or alternatively only an open lower limit $n > n_*$, there will not be an optimal program; for any feasible program can always be improved upon by running the population down a little faster. It is a simple extension of this result to show that if also $L(0) \geq 0$ is controllable and n is not constrained, so in effect L_t is controllable for all $t \geq 0$, an optimal program exists and has the property $L(t) \equiv 0$ for all $t \geq 0$. The optimal program is that there should exist no one; this is the goal that 'society' would wish to achieve.

Apart from an obvious inconsistency here the point is that in the absence of rather fortuitious circumstances such as population externalities in the utility function and/or strong economies of scale (more specifically, an increasing average product of labor with a constant capital stock) any program can be improved upon by running population down a little faster. This seems to be a compelling reason for rejecting this particular average utilitarian approach, as represented by the welfare function A, even if such a function is considered a priori ethically acceptable.

Pitchford [31] has considered a model in which the economy must be directed from an initial point (K_0, L_0) to a given terminal point $(\hat{K}, \hat{L})$ at some unspecified time T so as to maximise the welfare function,

$$W = \int_0^T [u(c) - u(\hat{c})]dt$$

The form of the technology $Q = F(K,L)$ is specified so as to yield a unique interior point $(\hat{K}, \hat{L})$ corresponding to a maximal level $\hat{c}$ of sustainable per capita consumption. This implies that, evaluated at $(\hat{K}, \hat{L})$:

$$F(K,L)/L \equiv q = \hat{c}$$

$$q = F_L \text{ and } F_K = 0$$

So, for the solution $(\hat{K}, \hat{L})$ to be unique and maximising, a typical K-cross-section of the production function must have the form illustrated in Figure 1. There is therefore an inherent element of non-convexity in the technology. However, by specifying a sufficiently large subsistence level $\underline{c}$ of per-capita consumption a feasible program will not locate in this non-convex region.

In order to use this model to examine the economic basis of expenditures on changing the population growth rate through birth subsidies, immigration and emigration it is specified that:

$$\dot{L}/L = \delta + \alpha i - \beta j$$

where i(j) is the per-capita expenditure on population increasing (decreasing) policies. These expenditures must be sustained to have a sustained effect on the rate of growth of the population.

Some comments on the formulation of this model are in order. There is no discounting of future utilities so, following Ramsey [33], the welfare function is bounded by defining it in terms of accumulated deviations of per-capita utility from $u(\hat{c})$; in this manner the problem of the possible non-existence of an optimal solution can be avoided.

Also observe that the economy is being directed to the point $(\hat{K},\hat{L})$ which is specified a priori rather than flowing from considerations of optimality. This terminal state is none other than the Cannan-Wicksell-Wolfe stationary state corresponding to $K = \hat{K}$ i.e. $\hat{L} = L_w = L_c$ and $F_K(\hat{K},\hat{L}) = 0$.

Finally, notice that it is a requirement of the model, rather than an implication, that the terminal population size be positive and furthermore there is now a direct cost necessitated by any program that reduces population size. One suspects, then, that due to these constraints the unpalatable implications (discussed earlier) of this social welfare function may not be observed. So it is of interest that there will be a finite period of time immediately prior to achievement of the terminal state in which, irrespective of the initial state of the economy, the optimal program will call for expenditures on reducing the population growth rate even if $\delta = 0$. In fact if the utility function is linear then all surplus output (over and above that required to maintain per-capita consumption at its subsistence level) will be so allocated, i.e. the population is run down at the maximal feasible rate.

Sato and Davis [36] have considered the problem of maximising the social welfare function W_A subject to a constant returns to scale technology so the constraints are given by equations (1) and (2) above. However, it is now assumed that population growth depends on per-capita income; i.e. $\dot{L}/L = n[f(k)]$. Equation (1) may be re-written:

$$\dot{k} = x(k) - c$$

where $x(k) \equiv f(k) - kn[f(k)]$ is 'net' output per capita. The Pontryagin necessary conditions yield:

$$p = u'(c)$$

$$\dot{p}/p = \rho - x'(k) \; .$$

If the population growth rate is constant then $x(k)$ is a strictly concave function with a unique maximum; this point corresponds to the 'golden rule' solution at which sustainable per-capita consumption is maximised. But if population growth is endogenous no reasonable assumptions on the form of $n(\cdot)$ will determine the form of $x(\cdot)$. So in general there is an inherent element of non-convexity in the 'net' feasible production set and this is the sole qualitative difference between this model and that in which population growth is exogenous [23]. Sato and Davis assumed $x''(k) < 0$ at a stationary point, thereby not appreciating the remark just made, and not surprisingly conclude that the optimal program has the same qualitative feature as in the exogenous population case.

A correct analysis [23] allows for the possibility that $x''(k)$ is not always negative so that there may be multiple k-solutions to the equation $\dot{p} = 0$ and multiple 'local' golden rule points. The optimal program can be shown to approach a stationary point and these points are alternatively saddle points or totally unstable. The situation is illustrated in Figure 2.

One notices that for any initial capital-labor ratio there will be more than one program which is a candidate for optimality (dotted lines). Also the configuration of the Pontryagin programmes is not uniquely determined when $x''(k)$ is not one-signed; and per-capita consumption may not be monotonic along the optimal program. These features are a direct implication of the inherent element of non-convexity induced by endogenous population change.

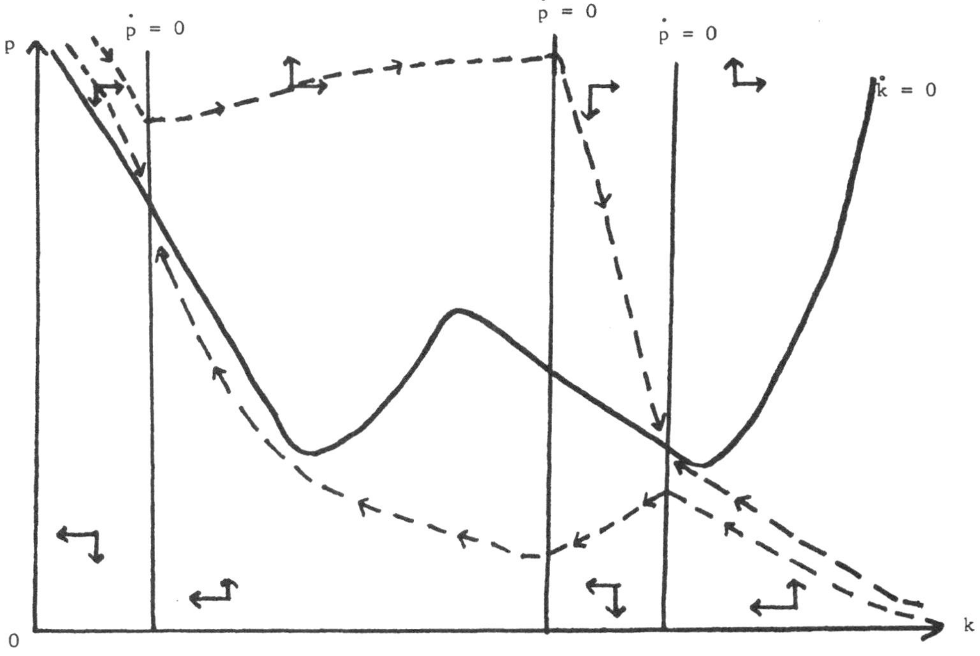

Figure 2.

Existence proofs have typically needed to utilize convexity assumptions. However, for a large class of models, recent results by Chichilnisky [3] have shown that convexity is not required. So suppose an optimal programme exists. Also let $p = p^T(k)$ be the optimal programme for a finite horizon T. Define:

$$p = \lim_{T \to \infty} p^T(k) = p^\infty(k).$$

If this limit exists, so $p^\infty(k)$ is defined, then Ruff [35] has given a heuristic proof of the proposition that it cannot be dominated by any other programme and so is optimal. This proposition is certainly valid under conditions of convexity and Ruff's contribution was to relax these conditions.

Now T-optimality requires $k(T) = 0$; for such a programme must satisfy the finite horizon transversality conditions. If $p = p_1(k)$ and $p = p_2(k)$ are the two candidates for optimality in Figure 2, and $p_1(k) > p_2(k)$ for all k, then:

$$p^\infty(k) \equiv p_2(k)$$

so this argument suggests that $p = p_2(k)$ is the optimal programme i.e. the optimal programme is uniquely identified as the Pontryagin programme which maximizes initial per-capita consumption subject to the constraint that it is feasible for infinite time. Notice that there is not a threshold theory of growth even though there are multiple stationary points.

One can show that:

$$q_t \equiv \partial W_A / \partial L_t < 0$$

along any program satisfying the necessary conditions i.e. if resources can be allocated to the direct control of the population growth rate then they should always be such as to reduce this rate. This adds force to the remarks made earlier; namely that with social welfare function A it is always optimal, given that it is feasible, to run down the population a little faster.

Earlier this social welfare function was noted not to have the property of symmetry of treatment over time; it provides a framework which is totally insensitive to distributional considerations. As a result social welfare function B was proposed:

$$W_B = \int_0^T e^{-\rho t} [L_t u(c_t) / \int_0^T L_\tau d\tau] dt. \quad \rho \geq 0$$

It is now appropriate to consider the a priori ethical foundation of such a welfare function.

Suppose that each moment of time is identified with a unique generation within which income is distributed in an egalitarian manner amongst the identical individuals. Now consider how to determine a "just" distribution of income across generations. Rawls [34], extending an argument of Harsanyi [14] and Vickrey [42], argues that if all members of the "original position" are ignorant of which generation they will belong to then they would each identify this just distribution as that implied by the maximisation of social welfare function B; for this is equivalent to maximising each member's expected utility where $L_t / \int_o^T L_\tau d_\tau$ is the probability of being a member of generation t and so receiving $e^{-\rho t} u(c_t)$. This line of reasoning is the "original position" argument of social contract theory where individuals must propose laws for governing society without knowledge of their personal identity. It is applied in an intertemporal context by Rawls to determine a just intertemporal consumption profile corresponding to a population profile which is given and so not subject to change.

This expected utility maximisation approach is used by Cooter [4] in an attempt to determine the optimal rate of population growth. Unfortunately the results of the analysis are quite incomplete and so will not be discussed. But whether the above line of reasoning is legitimate when the population profile is itself to be determined by someone in the "original position" is certainly dubious; in fact, in Section 5 below, it will be argued (following Dasgupta [9]) that it is not legitimate.

In contrast to social welfare function A, and also the classical utilitarian version used in sections 3 and 4, social welfare function B is inter-temporally inconsistent [4]. Specifically if generation g re-computes the optimal program, i.e. chooses it to maximise

$$\int_g^T \frac{L_t}{\int_g^T L_\tau d_\tau} \quad u(c_t)dt \ .$$

subject to $K(g) = K^*(g)$ where $K^*(g)$ is the capital stock bequeathed to generation g as a result of the program put into effect by generation zero, then the two programs will not agree for all g from time g onwards. But our line of reasoning requires generation g to be a member of the "original position" applicable not only at time g but also at time zero. So our reasoning requires that everyone in the "original position" at time zero must assume that the ethical ideal will be upheld even though they know this to be false; actual behavior is inconsistent with the assumptions under which the social welfare function is determined. Therefore there is an internal contradiction inherent in the line of reasoning used to justify this social welfare function.

Now suppose that every generation g computes the optimal program from time g onwards but puts the program into effect for the first time period only. Then each generation is deciding how large the next generation should be if procreation and savings are fair to future generations i.e. each generation is deciding its own responsibilities without supposing that past or future generations will live up to theirs. The problem of intertemporal inconsistency, and so also the internal contradiction stated above, then dissolve for it is only instantaneous procreation rates rather than a complete schedule which must be determined [4]. In summary; even if the "original position" argument is deemed applicable to problems of determining an intertemporal population profile social welfare function B can be justified only under conditions of continual planning revision.

Morishima [29] considers a close cousin of social welfare function B in the context of a discrete-time model with multiple commodities in which population grows at a rate determined by the per-capita consumption of these commodities. The technology is then viewed as transforming 'social inputs' into 'social outputs'; the inputs and outputs being both of commodities and people. The set of such transformations, which are point-to-set correspondences because the output of commodities depends not only on the input of commodities and people but also on the growth of the population during the period, i.e., on the goods allocated to their consumption, is assumed to remain constant. This production possibility set is required to have various properties but in particular it is a closed convex cone in Euclidean space. It therefore admits of balanced growth paths and the one that has the highest growth rate, 'α' say, is referred to as the turnpike which is assumed to be unique.

Alternative feasible growth paths must be compared. This requires specification of the social welfare function and it is defined to be:

$$W = W\left(\{\sum_{t=0}^{T} \theta^{-t} L_t u(c_{it})/T+1\},\ \sum_{t=0}^{T} \theta^{-t} L_t/T+1\right)$$

where c_{it} is per-capita consumption of commodity i in period t,

$$u(c_{it}) = c_{it}$$

and θ is the subjective time preference factor. This function is assumed to have the usual properties (of continuity, etc.) but in particular must:

(i) reveal the turnpike to be preferred to a 'state of nothing'

(ii) satisfy the postulate of quasi-homogeneity relative to the turnpike.

Condition (ii) together with $\alpha \geq \theta \geq 1$ then imparts the following normative property to the turnpike:

> among all possible balanced growth paths the turnpike yields a maximum to the given social welfare function.

Morishima then shows that if the programming period T is sufficiently long any feasible welfare maximising program will have certain well-defined properties of convergence to the turnpike.

As the results are strictly asymtotic nothing can be inferred about an optimal savings/population policy in the intermediate states en route to the turnpike unless the turnpike and the optimal growth path coincide, i.e. unless the intial conditions lie on the turnpike. Also observe that in a one-commodity world if $\theta = 1$ (so there is no discounting of future utilities) and if condition (ii) is strengthened to homogeneity of degree zero (with respect to $\sum_{t=0}^{T} \theta^{-t} L_t / T+1$) then this social welfare function is identical to social welfare function B with a linear utility function.

In the above presentation we have implicitly assumed full employment of the labor force; as a result, it has not been necessary to distinguish between the size of the labor force and of the population. In Morishima's analysis such a distinction is made. Furthermore, as the form of the population growth function is not specified, it is necessary for the model to be consistent with the existence of unemployment. Even along the turnpike there may be unemployment of labor; both the growth rate of the economy and of the population are mutually interdependent, in contrast to the von Neumann model, but are not necessarily the same. As this particular interaction is not examined, and as only asymtotic results are derived, the population growth function need not be specified. But a specification would be required if one hoped to say anything about the intermediate states enroute to the turnpike.

In the next two sections a classical utilitarian social welfare function will be postulated. It will be shown not to lead to unpalatable implications and furthermore it will be possible to derive implications for a joint optimal population/savings program even along the intermediate states.

3. CLASSICAL UTILITARIANISM: first best solutions.

Sidgwick [38] considered that if the average happiness of human beings is a positive quantity, then, assuming the average level of happiness will not decrease, utilitarianism directs us to make the number enjoying it as large as possible. But if this increase is at the expense of a decrease in average happiness, and if utilitarianism prescribes happiness as a whole to be the ultimate end of action rather than any individual's happiness, then the amount of happiness gained by the extra numbers should be weighed against that lost by the remainder and vice-versa for a decrease in numbers. So utilitarianism prescribes that the population should be of a size that maximises, not average happiness $u(c)$, but total happiness $Lu(c)$ i.e. the social welfare function, in a static context, is:

$$W = Lu(c)$$

Writing almost a half century later Meade [27] revived and extended this formulation. In the Sidgwick-Meade formulation the optimal size of the population L_M is such that to maximise the welfare function $Lu(c)$ subject to the technology $Q = F(K,L)$ where both the capital stock and the level of savings are fixed. Then L_M must satisfy:

$$\text{(M)} \qquad u(c) = u'(c)\{c - F_L\}$$

Now $c-F_L$, evaluated at $L = L_w$, is zero and increasing in L so clearly $L_M > L_w$, the Wolfe optimal population size; and at this optimum the marginal product of labor is less than average consumption. Meade's rule (M) concerns the desirability, given that an optimal savings program has been determined, of introducing an additional member into society; the 'benefits' and 'costs' of such an action should be balanced. The benefit to society, in the event that consumption is distributed equally, is the level of welfare of that individual, i.e., u(c). The cost to society is the excess of this individual's consumption over his marginal product and evaluated at the correct scarcity price u'(c) which includes the direct effect of this individual on the social welfare function.

This formulation requires that one compare the state of being alive with the unborn state so if one imputes, simply as a normalization, a zero level of utility to the unborn state, the sign of the utility level becomes a matter of great importance in considerations of population policy. Also Meade's rule indicates that one must specify the level of per-capita consumption c_o at which average utility $u(c_o)$ is zero. Thus one is being asked to make some evaluation as to the economic opportunity or the quality of life necessary to justify bringing an additional child into society i.e. c_o. In both the classical and average utilitarian approaches one must consider whether or not to bring an additional child into society; but only in the classical approach is the above consideration given explicit recognition. This is, surely, a strength of the classical theory even though one can doubt whether exact specifications of c_o is ever possible. In fact the models below show, at least from the point of view of the qualitative characteristics of an optimal savings population program, that the required comparison necessitates specification of c_o only in so far as it is in some sense 'high' or 'low'; and this is certainly the kind of judgement that people have always considered.

The natural inter-temporal extension of this social welfare function is:

$$\text{(C)} \qquad W_c = \int_o^\infty e^{-\rho t} L_t u(c_t) dt \qquad \rho \geq 0$$

Consideration of our earlier two island example (p.8) indicates that the classical social welfare function requires per-capita consumption to be equalised among all members of society irrespective of which island one locates on. Therefore social welfare function (C) has the desirable property of symmetry of treatment over time, i.e. it is sensitive to distributional considerations. In addition it is intertemporally consistent and furthermore the principles on which it is based are explicit and do not admit of internal logical contradictions. It therefore seems to be a marked improvement over social welfare functions (A) and (B) and so is used for the remainder of this section and also section 4.

The first best problem of choosing a savings population program that maximizes W_c, when the population size is instantaneously and costlessly controllable at every moment of time including the present, has been considered by Dasgupta [6].

Suppose there is a fixed factor of production which causes diseconomies of scale. This assumption is implicit if a Cobb-Douglas production function is postulated with the sum of the exponents less than unity, i.e.

$$Q = F(K,L) = K^{\alpha}L^{\beta} \qquad \alpha,\beta > 0 \qquad \alpha+\beta < 1$$

The utility function u(o) is increasing, strictly concave, twice differentiable, $u'(o) = \infty$ and $u'(\infty) = 0$. Recall that K(0) is given but L(0) controllable and suppose also positive discounting of future utilities i.e. $\rho > 0$. The constraints are:

$$\text{(C)} \qquad \begin{aligned} \dot{K} &= F(K,L) - Lc \\ C,L,K &\geq 0 \end{aligned}$$

where the time index is subsumed.

The Pontryagin necessary conditions yield:

$$\text{(M)} \qquad u(c) = u'(c)[c-F_L]$$

$$\text{(R)} \qquad \dot{p}/p = \rho - F_K \text{ where } p = u'(c)$$

which are sufficient if:

$$\text{(T)} \qquad \lim_{t\to\infty} e^{-\rho t}pK = 0$$

The first condition is the Meade rule discussed earlier. The second condition is Ramsey's result; namely that the percentage rate of fall of the marginal utility of consumption should be equal to the 'net' marginal product of capital i.e. $F_K - \rho$.

The sufficiency condition (T), commonly referred to as the transversality condition, requires that the present value of capital in the distant 'future' tends to zero as the 'future' recedes.

A stationary solution $(\underline{c},\underline{K},\underline{L})$, so $\dot{\underline{c}} = \dot{\underline{K}} = \dot{\underline{L}} = 0$, to the above system exists if the utility function has a positive least upper bound. It will be characterized by

$$\underline{q} \equiv \underline{F}/\underline{L} = \underline{c}$$

$$F_{\underline{K}} = \rho$$

$$u(\underline{c}) = u'(\underline{c})[\underline{c} - F_{\underline{L}}]$$

$$= \underline{c}u'(\underline{c})(1-\beta) > 0 \text{ as } F = K^{\alpha}L^{\beta} \text{ and } 0 < \beta < 1$$

A unique optimal program will exist and asymtotically it will approach this long run steady state $(\underline{c},\underline{K},\underline{L})$. If $K(0) = \underline{K}$ the optimal program is $(\underline{c},\underline{K},\underline{L})$. If $K(0) < \underline{K}$ then along the optimal program c,K and L are monotonically increasing with time and vice-versa.

Notice that if K(0) is sufficiently small then there is a finite period of time during which welfare u(c) is negative and then welfare becomes positive and increases to its asymtotic value. Nevertheless population is increasing all the time i.e. people are being 'brought in' even when welfare is below the subsistence level. The reason is that it is the overall welfare of generations that is being maximised; so when the initial capital stock is low people are allocated consumption rates less than their marginal products so as to accumulate capital in the interests of posterity.

If $u(c) = B - c^{-v}$ $(B, v > 0)$ then one can deduce that $\underline{c}/c_o$, where $u(c_o) = 0$, is less than 'e' i.e. 2.74. If also $\beta = 0.5$ and $v = 1$ then $\underline{c}/c_o = 3/2$. It seems, therefore, that in a world that has a fixed factor of production the long run standard of living on the optimal program is not much larger than the welfare subsistence rate c_o - no matter how the latter is defined.

If there is no discounting of future utilities, $\rho = 0$, then the optimality conditions are inconsistent with any steady state; it appears, then that in this case an optimal program will not exist.

Now suppose $F(K,L)$ is linear homogeneous so, by definition, $f(k) \equiv F(k,1)$. Assume $f(\cdot)$ to be increasing, strictly concave, twice differentiable and to satisfy the boundary conditions $f'(0) = \infty$ and $f'(\infty) = f(0) = 0$. Retain the assumption $\rho > 0$ and also the previous assumptions on $u(\cdot)$ but, rather than specifying a positive least upper bound on $u(\cdot)$, require that $\lim_{c \to 0} u(c)/u'(c) = 0$ (so in particular all constant elasticity utility functions are admisable). Define k^* to be the unique solution to $f'(k) = \rho$ and if $c_o > 0$ (where $u(c_o) = 0$) define c^* to be the unique solution to $\partial Lu(C/L)/\partial L = u(c) - cu'(c) = 0$. Clearly $c^* > c_o$. Observe that $K(0)$ is given and $L(0)$ is controllable so $k(0)$ is controllable.

The constraints (C) are as above and the Pontryagin conditions yield rules (M), (R) and (T) but now $F_L = f(k) - kf'(k)$ and $F_K = f'(k)$. Analysing these conditions [6 or 25] indicates that:

(i) If $c_o \leq f(k^*) - k^*f'(k^*)$ then no optimal program exists.

(ii) If $c_o > f(k^*) - k^*f'(k^*)$ then there exists a steady state optimal program with $k = k^*$, $c^* > c > c_o$ and $r \equiv \dot{L}/L < \rho$.

Earlier it was stated that the welfare subsistence rate 'c_o' must be specified to be 'large' or 'small' and results (i) and (ii) give this a precise meaning. The results state than an optimal program exists if and only if 'c_o' is not too 'small'; specifically, it must be greater than the marginal product of labor at that point where the 'net' marginal product of capital is zero. When an optimal program does not exist the reason is intuitively clear. Restricting our attention to steady state paths, for otherwise the economy exhausts its capital stock in finite time, a sufficiently small 'c_o' means it is possible to have a rate of growth of the population in excess of 'ρ' and still maintain positive utility. Therefore any given program can be improved by keeping consumption low (and utility non-negative) for a finite initial period and then enjoying a splurge in the distant future when there will be more people to enjoy the splurge. But social welfare can always be increased by postponing the splurge so an optimal program does not exist. In contrast, if 'c_o' is large enough, then high population growth rates can be achieved only at the expense of positive utility; in this case a balance will be made between the two. If 'c_o' takes the borderline value $f(k^*) - k^*f'(k^*)$ then total welfare is zero and it is clear that a small increase in per-capita consumption, even at the expense of a smaller population, will be an improvement and so no optimal program exists.

It is a simple extension of these results to show that if there is no discounting of future utilities, and the utility function has a positive least upper bound B say, then there is no steady state solution to the Meade and Ramsey rules, i.e. all Pontryagin programs exhaust their capital stock in finite time so no optimal program exists. Suppose capital saturation is introduced so near $k = \hat{k}$ say strong diminishing returns set in, specifically, $f'(k) = 0$ for all $k \geq \hat{k}$. Write $\hat{c} = f(\hat{k})$. Then if and only if 'B' is small enough, i.e. 'c_0' high enough, will an optimal program exist. It will be unique and will be characterized by per-capita consumption in excess of $\hat{c}$ and a negative population growth rate.

For comparative purposes recall that under average utilitarianism the optimal program for the above economy requires $L_t \equiv 0$ for all $t \geq 0$. Also if $L(0)$ is given but $r \geq -m$, $m > 0$, is controllable then the optimal program would have called for setting $r = -m$. Under classical utilitarianism, however, and with positive discounting, 'r' takes on its lower bound value only when $k(0) < k^*$; and in this case the Ramsey rule applies (but not the Meade rule) and per-capita consumption increases monotonically. If, on the other hand, $k(0) > k^*$ then 'r' takes on its upper bound value $M > \rho > 0$ say; again the Ramsey rule applies but now per-capita consumption decreases monotonically. Such a program is followed until $k = k^*$ at which point the Ramsey and Meade rules are both followed and results (i) and (ii) above will apply.

If Harrod-neutral technical progress is introduced into either of the technologies considered above then later generations are assured high enough consumption rates even if population growth rates are high. Consequently large discount rates are required to ensure the existence of an optimum; in fact if the technology is of the constant returns to scale variety and $f'(0) = \infty$ then no optimum exists for any finite discount rate. This is in sharp contrast to the results applicable when population grows at an exogenously given rate and the introduction of technical change weakens the conditions required for existence.

Suppose population externalities due to congestion are introduced into the individual's utility function e.g. by specifying,

$$u(c) = B(L) - c^{-v} \qquad v > 0$$

where B(L) is no longer a constant but:

$$B'(L) < 0 \quad B''(L) < 0 \text{ and for some } L_o,\ B(L_o) = 0.$$

If the technology is of the Cobb-Douglas decreasing returns to scale variety then it is intuitively clear that the long run population level will be smaller than if B had been a constant at B(0). Also long run consumption per head will be higher and the capital stock lower. Additional properties of the optimal program will be as before. But suppose now the technology exhibits constant returns to scale. Then the introduction of population externalities makes the model more akin to the one incorporating a decreasing returns to scale technology. The optimal program, in general, will no longer be a steady state; but rather it will approach the steady state asymtotically if one exists.

There is a certain dichotomy between the savings/population decisions in all the models considered in this section; it is represented by the Meade and Ramsey rules not being interdependent. This is a consequence of the controllability of L(t). In the context of a constant return to scale technology, and if L(O) itself in controllable, the model is not fully dynamic; for it is always possible to choose the initial state of the system, i. k(O), such that the asymtotic state is achieved immediately. There can be no analysis of the optimal transition to this state. In a model where there is some cost to controlling the population size, either direct or indirect, an analysis of this transition will be necessary and furthermore the dichotomy will not appear; the Ramsey and Meade rules will be interdependent. Such an analysis is considered in section 4 below.

The results of this section are certainly of normative interest; they indicate what society would like to do if given complete control over the state of the economy. The optimal program will clearly be preferable to that given when such control is not possible. Now if the technology exhibits constant returns to scale then results (i) and (ii) above, and also the extension of Ramsey's analysis of an optimal intertemporal savings program (with $W = W_c$) to include positive subjective time preference together with population growth at a constant rate, both state that if $k(t) > k^*$ (where $f'(k^*) = \rho$) the optimal program calls for a reduction in $k(t)$ and vice-versa. This suggests:

an 'over-developed' economy is one for which $k(0) > k^*$ and
a 'less-developed' economy is one for which $k(0) < k^*$.

One of the purposes in the next section is to determine when an optimal program will transfer from a less-developed region to an over-developed region and vice-versa.

4. CLASSICAL UTILITARIANISM: Second Best Solutions.

The explanation for the increased world population growth rate, referred to in section 1, is commonly called the theory of the demographic transition [15,16]. In essence this theory states that the high growth rates have resulted from an increased gap between mortality and fertility rates; in turn this has been due primarily to reductions in the former. The major causes of reduced mortality rates are the development of modern medicine and public health programs, improvement in the standard of living and improved environmental sanitation. In the western world most of the reduction in the mortality rate has occured over the last 100 years and has been followed by a decline in fertility; but with a lag and not to the same extent. In the non-western nations the mortality rate decline has taken place over the last 25 years and so far only in Japan and mainland China has this been followed by a significant decline in fertility rates. In societies that suffered from high mortality rates in the recent past the existence of high fertility rates can be readily appreciated due to certain socio-cultural factors such as the need to ensure sufficient survivors to carry on the family name and responsibilities. In short, cultural attitudes appropriate to the new environmental condition of low mortality rates have not yet been developed. Therefore this theory of the demographic transition explains the persistence of high population growth rates in many countries together with higher growth rates in the developing as compared to the developed nations.

The fertility rate is the major determinant of changes in the population growth rate in any given country. The question naturally arises as to whether economic development, in the sense of an increase in per capita income for example, enhances or inhibits fertility. The theory of the demographic transition suggests that fertility rates decline as development proceeds, providing that the transition is being realised. Thus Kingsley Davis [19] for instance, groups countries into three classes. Class 1 pertains to nations that are highly developed, have low fertility and mortality rates and a very low population growth rate.

Class 2 pertains to those nations that are beginning to develop, have a high fertility rate and low mortality rate although both are declining, and a high overall population growth rate. Classes 1 and 2 correspond to nations in which the demographic transition is in process. In contrast Class 3 nations have not begun to develop and have both high fertility and mortality rates and a moderately low population growth rate. Malthus, however, believed that development leads to an increased demand for labor which in turn encouraged earlier marriage which then leads to an increase in the fertility rate [11].

Both Heer [17], using the conceptual framework of Spengler [39], and also Becker [2] have provided a reconciliation of these opposing views. In the process an explanation of the differential fertility rates mentioned earlier is provided. The decision to have or not to have an additional child is assumed to be the result of a rational decision making process; namely that of the economic theory of consumer demand for a durable commodity. Therefore the outcome will be determined by the interaction of the level of income, the price system and the preference structure where all three elements must be given a suitably broad interpretation; for example, the preference structure delineates the relative value associated with different goals, both economic and non-economic.

By analogy with consumer demand theory, if income is rising it will be more likely that a decision will be made in favor of having the child (direct effect); so if, in fact, the demand for children is decreasing the cause must be found in either a change in the price system or the preference structure or both (indirect effects). The substance of Heer's thesis is that the indirect effects may outweigh the direct effects; so reconciling the Malthusian viewpoint with the theory of the demographic transition.

It behooves us to consider, therefore, the manner in which the price system and the preference structure may have changed. The relative price of children has increased for several reasons, for example (1) due to the development of new and improved methods of birth control the cost of not having a child has decreased, (2) development has led to urbanisation which leads to a rise in the relative cost of living space; in turn the demand for living space is complementary with the demand for children, (3) the labor cost of child care relative to that of producing material goods has increased; this is especially important in light of the growing opportunities for women in the market economy and, (4) the cost of children is higher in developed countries where the requirements of a complex technology and of social norms require a longer period of education for the child and so a longer period of child dependency for the mother. Changes in preferences accompanying development have also led to reduced fertility. The reduction in the infant mortality rate has reduced the desire for children because fewer births are required to ensure a given number of survivors. Also the value of a child's productive labor to the parents has been reduced through the introduction of compulsory education and of social security schemes; with the latter it becomes unnecessary to have many children in order to provide support in old age. In addition there has been a decline in the social status of having many children; at least in part because various governments and religious bodies no longer encourage large families. Moreover, economic development usually shifts the criteria for social status towards achievement;

thereby sometimes leading to a desire for conspicuous consumption and a consequent reduction in the relative value of children.

It appears reasonable, then, to assume that the population growth rate is an increasing function of per-capita income at sufficiently low levels of per-capita income and thereafter decreasing. In the remainder of this section the implications of classical utilitarianism, for the choice of an optimal savings program subject to the constraint imposed by endogenous population change, are examined [22]. Initially it is assumed that:

$$\dot{L}/L = n[f(k)] \qquad n'(\cdot) > 0$$

where f(k) is per-capita income i.e. the direct effects dominate. This is later modified in such a way as to consider the implications of the indirect effects dominating ($n'(\cdot) < 0$), the introduction of an explicit program that alters the population growth rate via the re-allocation of resources to this end, and also the change in the age structure of the population implied by changes in its rate of growth.

(1) <u>A Neo-classical Technology</u>

Assuming constant returns to scale, and the absence of population externalities such as congestion and pollution, the problem can be immediately stated as that of finding the per-capita consumption programme that maximises

$$W_B = \int_0^\infty e^{-\rho t} L\, u(c)dt \qquad \rho > 0$$

subject to the constraints imposed by the technology, the endogenous population growth rate and feasibility,

$$dk/dt = f(k) - kn[f(k)] - c$$

$$c \geq 0,\ k \geq 0,\ K(0) \text{ and } L(0) \text{ given}$$

where the dependence of all variables on time is implicit.

Such a model has two qualitative features which make it essentially different from its counterpart with exogenous population growth. These are (i) the need to specify the welfare subsistence rate c_o at which $u(c_o) = 0$ due to the non-invariance of the welfare function under linear increasing transformations of the utility index and (ii) the possible non-convexity of the 'net' feasible production set (as defined by $f(k) - kn[f(k)]$) and also the welf function (due to the dependence of L on k). Whilst Davis and Sato [36] have considered this model their analysis does not incorporate these points; an evaluation of their work appears in [22] and a correct analysis of this model in [23].

In what follows we will write $r(k) \equiv n[f(k)]$ and L(0) will be normalized equal to one. Now:

$$W_B = \int_0^\infty e^{-[\rho t - \int_0^t r(k(s))ds]} u(c)dt$$

so our model is now equivalent to a one state variable problem but with a variable discount rate, i.e., it is non autonomous. An analogous situation was encountered by Uzawa [41] in his study of the relationship between endogenous time preference rates and optimum asset holdings; however, by suitably transforming the units of time it was possible to make the model autonomous. We can use the same approach if we assume:

there exists an $\varepsilon > 0$ such that, for all k,

$n[f(k)] \equiv r(k) \leq \rho - \varepsilon$.

Furthermore this assumption ensures that the social welfare function converges and this avoids existence problems. In particular the infinitely deferred splurge, referred to earlier, cannot arise.

Now define $\Delta = \rho t - \int_o^t r(k(s))ds$. This is a strictly increasing transformation of t and ranges from o to ∞ when t ranges from o to ∞. Therefore the problem reduces to finding a per-capita consumption program that maximizes:

$$(1) \qquad W_B = \int_o^{\infty} e^{-\Delta} u(c)/[\rho - r(k)] \cdot d\Delta$$

subject to

$$(2) \qquad \dot{k} = [f(k) - kr(k) - c]/[\rho - r(k)]$$

$$(3) \qquad k, c \geq 0, k(0) \text{ given}$$

where all variables are implicitly functions of the new 'time' index Δ and the notation $\dot{k}$ is used to denote differentiation with respect to this new 'time' index. Clearly the model is now time-autonomous i.e. the optimal policy will not depend explicitly on Δ. It will be assumed that $f(\cdot)$, $u(\cdot)$, $n(\cdot)$ and so $r(\cdot)$ are increasing, strictly concave and twice differentiable, that $f(\cdot)$ satisfies the Inada derivative conditions and that $u(\cdot)$ satisfies $\lim_{c\to o} u'(c) = \infty$, $\lim_{c\to\infty} u'(c) = 0$. Recall that k^* is defined such that $f'(k^*) = \rho$ and c^* such that $u(c^*)-c^*u'(c^*) = 0$. Define also $p_o \equiv u'(c_o)$ and $p^* \equiv u'(c^*)$.

An application of Pontryagin's results [32] shows that there is a continuous function $p(\Delta)$ such that the optimal solution satisfies the following conditions:

(4) $u'(c) = p > 0$ so $c = c(p)$ say and $c > 0$

(5) $\dot{k} = \psi\,(p,k) \equiv [f(k) - kr(k) - c(p)]/[\rho - r(k)]$

(6) $\dot{p} = \Phi\,(p,k) \equiv \{p[\rho - f'(k)] - r'(k)q\}/[\rho - r(k)]$

where $q \equiv H - pk$ and $H(p,k) \equiv u(c(p))/[\rho - r(k)] + p\psi(p,k)$.

The term $q \equiv H - pk$ plays a fundamental role, as is suggested by Eq. (6), in the following analysis. It is therefore worthwhile to give some guide as to its interpretation. This is done in (a) and (b) below.

(a) The 'shadow price' of a unit of labor is q along Pontryagin paths that approach a stationary point i.e., if L(t) increases by one unit then the criterion increases by q(t) units. This is of direct interest as the optimal programme must approach such a point. It is intuitively clear, and easily proved, that the sign of q determines in which direction to seek to change population growth in the event that society has some direct control, via the explicit allocation of resources to this end, over the population growth rate.

(b) An alternative interpretation is provided by noticing that $q = 0$ is a necessary condition for the maximization of, where 'r' is the control variable,

$$\int_0^\infty e^{-\Delta}\, u(c)/[\rho - r]\, d\Delta$$

subject to $k(\Delta) = k(0)$ so 'c' is constrained by $c = f(k(0)) - rk(0)$. If in addition L(0) is controllable, as in some of the analysis of Dasgupta and Meade, the corresponding condition would be:

$$M(p,k) \equiv u(c) + u'(c)\,[f(k) - kf'(k) - c] = 0$$

where $p = u'(c)$. If $k = k^*$ then the two conditions are equivalent and, in fact, the optimal solution in Dasgupta's analysis does require $k = k^*$.

Clearly both $M(p,k)$ and $q(p,k)$ are related. In summary, both represent the 'shadow price' of a unit of labor in the respective models with which they are associated.

There are a few pertinent remarks that should be made concerning the interpretation of Eq. (6). As $dp/dt = [\rho - r(k)]\,\dot{p}$ (6) may be rewritten in the form:

$$(7) \qquad dp/dt = p[\rho - f'(k)] - r'(k)\,q$$

or

$$(8) \qquad -d[e^{-\rho t}\,p]/dt = [e^{-\rho t}\,p]\,f'(k) + [e^{-\rho t}\,q]\,r'(k)$$

Notice that the first part of (7) represents Ramsey's rule and reflects the direct role of capital accumulation. The second part reflects an interdependence, which does not arise when $L(0)$ is controllable, between the rules of both Meade and Ramsey.

Equation (8) indicates that (7) is an allocation rule determining the division of output between consumption and increases in the capital stock; it requires that the marginal unit of output be allocated in such a way as to equate the 'benefits' and 'costs.' Suppose the marginal unit is allocated to capital accumulation. Then the benefit is the marginal product of this unit in production, evaluated at the correct scarcity price $e^{-\rho t}p$, together with the shadow price of the change in population that it induces. This must be compared with the cost, i.e., the change in the discounted value of a marginal unit of consumption.

Notice that output is not being allocated between consumption, additions to the capital stock and population control. It is allocated only between the former two and the changes in population size are induced changes. The problem is to find an optimal savings policy subject to the 'constraint' imposed by endogenous population change. A true Ramsey-Meade integration, with a three-way allocation, would require the explicit introduction of government expenditures to control the size of the population.

An alternative interpretation is possible. The first part of (7) yields the manner in which perfectly altruistic individuals would plan their optimal consumption stream over time in a perfectly competitive environment where the rate of interest would be equal to the marginal product of capital. The second part of (7) then reflects the deviation between this form of behaviour and socially optimal behaviour that results from the inability of a perfectly competitive economy to correctly appropriate, back to the private citizen, the full social costs and benefits induced by the effect of his action on the future size of the population.

The method of determination of the optimal programme will only be sketched: full details having appeared in [24]. It should be added that the analysis in [24] proceeded under the additional assumptions $\lim_{c \to o} u(c)/u'(c) = 0$, $\lim_{k \to o} kr(k) = 0$ and $\lim_{k \to \infty} r(k) \geq \varepsilon > 0$. Whilst these assumptions tidy up the analysis they do not play an important role and so are not discussed here. The optimal programme can be identified as a solution to Eqs (4)-(6) that 'passes through' the points at which $\dot{k} = \dot{p} = 0$. These loci ($\psi = 0$ and $\Phi = 0$) are illustrated in the figures below, along with the loci $q = 0$ and $\psi = k$. The horizontal arrows indicate the direction in which k moves as Δ increases.

Notice that as a result of the non-convexity of the feasible production set, i.e., the non-concavity of $f(k) - kr(k)$, there may be multiple values of k for any given p such that $\psi = 0$. Nevertheless the location of the $\psi = 0$ locus is quite well specified by noting that it is bounded above by the locus $\psi = k$. (This latter is easily located relative to the $q = 0$ locus; in particular, they intersect, if at all, only when $p = p_o$).

The $q = 0$ locus enables us to determine the location of the $\Phi = 0$ locus. The form of the $q = 0$ locus is relatively simple because it is independent of $r(k)$. It is precisely this independence which facilitates examination of the optimal programme under alternative assumptions on $r(k)$. In addition the $q = 0$ locus determines the implications of the optimal programme for population policy.

Observe that the form of the $q = 0$ locusdepends critically on the sign of $c_o - [f(k^*) - k^*f'(k^*)]$. As 'q' is the scarity value of a unit of labor the results of Dasgupta's analysis pertaining to the case where L(0) is controllable indicate that this should be expected. In fact the programmes characterized by d_1, d_2 and d_3 in Figures 4, 5 and 6 represent the only programmes that satisfy the necessary conditions for the Dasgupta problem; and of these, only d_2 satisfies the sufficiency conditions. Therefore, when L(0) can be freely chosen, an optimal programme exists if and only if $c_o > f(k^*) - k^*f'(k^*)$, and it has $k(t) = k^*$ for all t. Recall that $k < k^*$ defines the 'lesser' developed region and vice-versa.

The form of the $\Phi = 0$ locus depends critically on the relative curvatures of r(k) and $f(k) - \rho k$, as summarised by the sign of

$$g = \frac{d}{dk}\left[\frac{r'(k)}{f'(k) - \rho}\right] .$$

Under the assumptions made so far the sign of $\Phi_k\big|_{\Phi=0}$ is the same as that of g; also $g > 0$ if $k \varepsilon (k^*, \infty)$ but, if $k \varepsilon (0, k^*)$, then the sign of g is indeterminate. However, this knowledge is required for the construction of the $\Phi = 0$ locus; for example, if g is not one signed there may exist multiple segments to the $\Phi = 0$ locus. The $\Phi = 0$ locus, as illustrated in Figures 3 to 5 inclusive, is based on the assumption $g > 0$ for all $k \varepsilon (0, k^*)$ and in Figure 6 on the assumption $g \geq 0$ for all $k \varepsilon (0, k^*)$. The vertical arrows indicate the direction in which p changes when Δ increases.

The above difficulty exists because of the inherent non-convexity of the model. The results of this sub-section relate only to the case where the non-convexities are 'small' in the specific sense that $\Phi_k\big|_{\Phi=0} \geq 0$ for all k (the motivation for this terminology is that this is necessary for the usual sufficiency results of optimal control theory [1] to be applicable). This assumption, which implies $g \geq 0$ if $r'(k) > 0$ and vice versa, is a definite restriction. It will include, for example, the case where r(k) is linear in the $(0, k^*)$ interval (and thereafter satisfies the assumptions made earlier) but it will exclude the case in which f(k) is representative of a constant coefficient technology. 'Smallness' of the non-convexities in the sense defined implies that for any given k(0) there is only one path that approaches a Pontryagin stationary point, i.e., only one candidate for optimality, something which is not true if ever $g < 0$.

Let $k_i (i = 1,\dots,n)$, $k_1 < \dots < k^* < \dots\dots < k_n$, be the capital-labor ratios corresponding to the Pontryagin stationary points. There will be an odd number of them and they will be alternatively saddle points or locally unstable. The solution to equations (4) - (6) which passes through a Pontryagin stationary point is illustrated by thick dashed lines in the diagrams below. It is easily shown to define continuous functions p(k), c(k), q(k) for all k.

The qualitative properties of the optimal programme, as depicted in Figures 3 to 6 inclusive, can now be discussed. First notice that the optimal programme determines a threshold theory of growth in which the asymtotic levels of 'c' and 'k' depend on k(0). This results from the possible non-convexity of the feasible 'net' production set and/or that of the social welfare function. As the combined effect of these non-convexities is small it is still possible to show that p(k) and c(k) are monotonic functions and also that $u(c(k_i))/[\rho - r(k_i)]$ is monotonic increasing in 'i'.

If $0 \leq c_o \leq f(k^*) - k^*f'(k^*)$ it is optimal to approach asymtotically a state of overdevelopment in which $k(\infty) > k^*$ and $u(c(\infty)) > 0$ even if the economy is initially in a state of lesser-development. If $c_o > f(k^*) - k^*f'(k^*)$ such an economy may be 'trapped' in a lesser developed state and

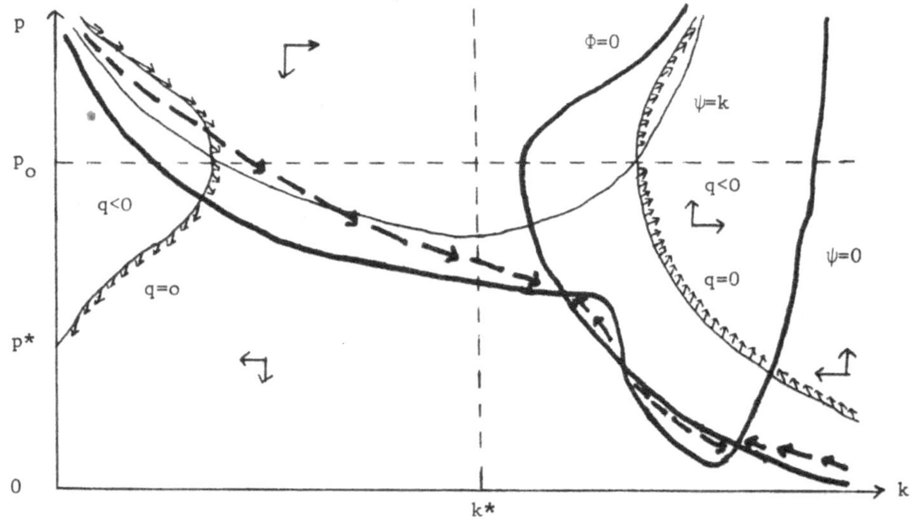

Figure 3: $c_o < f(k^*) - k^*f'(k^*)$

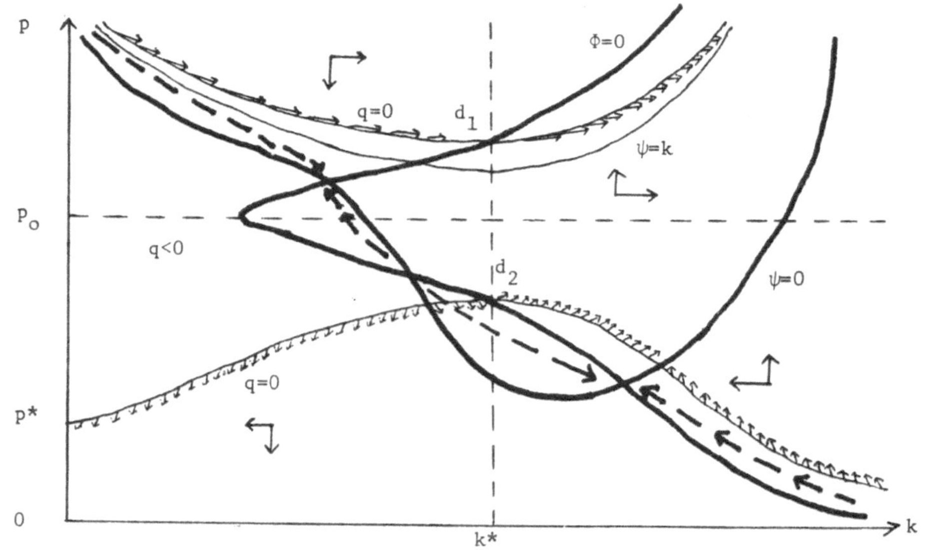

Figure 4: $c_o > f(k^*) - k^*f'(k^*)$

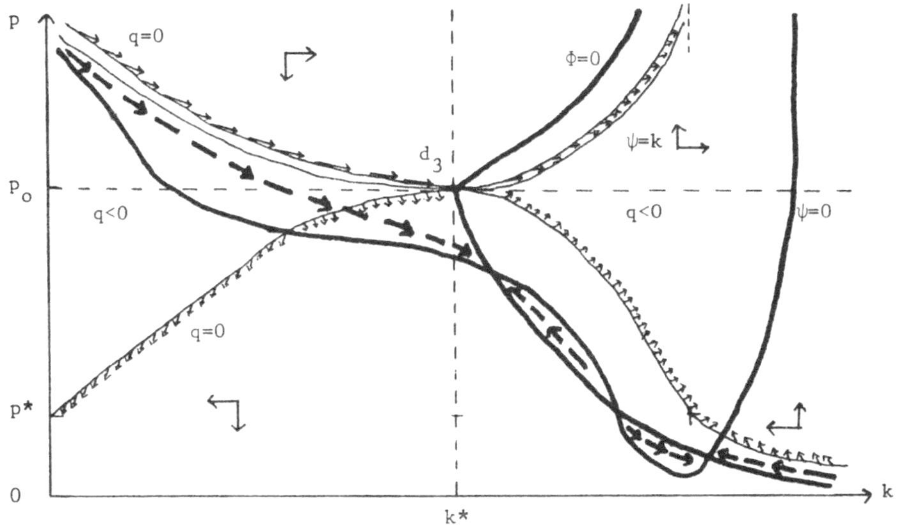

Figure 5: $c_o = f(k^*) - k^*f'(k^*)$

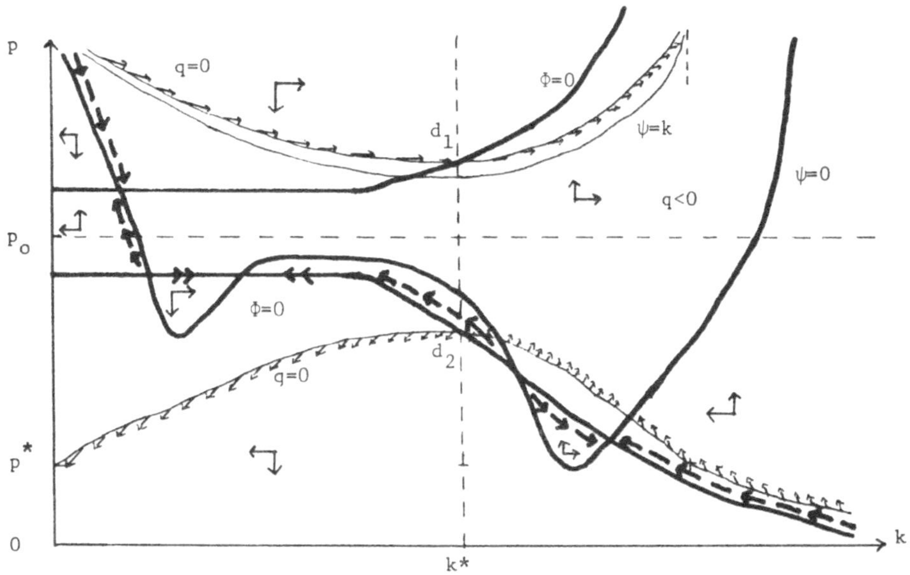

Figure 6: $c_o > f(k^*) - k^*f'(k^*)$

it is even possible that $u(c(\infty)) < 0$; in this case whilst it is feasible to overdevelop it is not optimal. The reason is that the cost of developing (arising in particular from the effect of L on c) is high due to the high c_o and outweighs the benefits (arising from the increased per-capita production made possible through a higher capital-labor ratio). Notice that there can be at most one asymtotic state with $u(c(\infty)) < 0$ and that necessary conditions for this are $k(\infty) < k^*$ and $c_o > f(k^*) - k^*f'(k^*)$.

Irrespective of the value of $c_o \geq 0$ if k(0) is sufficiently small and less than k_1 the optimal programme requires increasing the capital-labor ratio even though this increases the population growth rate (which may be negative) and is at the expense of per-capita consumption which is, and may always be, less than the welfare subsistence level. The explanation is that capital is being accumulated, or at least run down less fast than the population, in the interests of posterity; it is, after all, the over-all welfare of all generations which is being maximised.

Finally observe that if c_o is small, i.e. less than $f(k^*) - k^*f'(k^*)$, then it is never optimal to transfer from a state of over-development ($k > k^*$) to one of lesser-development ($k < k^*$). The strength of this result lies in its validity irrespective of the size of the non-convexities. In fact if $g < 0$ over some range there may exist possible asymtotic states in the region defined by $q < 0$ and $k < k^*$ so the result is not trivial. Its validity follows from consideration of the behaviour of Pontryagin paths in a neighborhood of the $q = 0$ locus (see the small arrows in the figures above), this being independent of the sign of g.

Now suppose there is an explicit population policy in the sense that resources can be allocated, in general at the expense of both consumption and investment, to the direct control of the population growth rate. Let 'm' be expenditure per-capita on such a policy and let 'v' take the value +1 if it is desired to increase the growth rate and -1 if it is desired to reduce the growth rate. Then:

$$\dot{L}/L = r(k,mv)$$

where $\partial(\dot{L}/L)/\partial mv|_k = r_2(k,mv) > 0$. Both 'm' and 'v' are control variables. Given the interpretation of q it is clear, and is proved in [25], that:

$$q = 0 \Rightarrow m = 0$$

$$m > 0 \Rightarrow qv > 0$$

i.e., if it is optimal to allocate some resources to the direct control of the population growth rate ($m > 0$) then the objective should be to reduce the growth rate ($v = -1$) if the shadow price of labor is negative ($q < 0$) and vice-versa. It is of interest to know how many times, and under what circumstances, v can switch sign along the optimal program.

Consider c and mv to be complementary if they are both moving in the same direction along the optimal path; if they move in opposite directions then they are substitutes. One would expect that whether these control variables are complements or substitutes would depend only on the signs of $d[q(k)/p(k)]$ and $p'(k)$, the former indicating the change in the relative price of K and L, providing:

$$r_{12} = r_{21} = 0$$

i.e., the direct effects on the population growth rate of changes in per-capita expenditure on population policy are independent. Furthermore, if $d[q(k)/p(k)]/dk$ is positive one expects dmv/dk to be positive, and vice-versa, providing:

$$v\, r_{22} < 0$$

i.e. there are diminishing returns to 'm' in its effect on the population growth rate. In fact [25],

$$v d[q(k)/p(k)]/dk = -[q(k)/p(k)]^2 \{r_{21} + r_{22}\, dmv/dk\}$$

so our assumptions imply,

$$d[q(k)/p(k)]/dk \gtreqless 0 \Leftrightarrow dmv/dk \gtreqless 0$$

as expected. It is of interest to know under what circumstances 'c' and 'mv' are complements or substitutes along the optimal programme.

Along the optimal programme one can show (i) $q'(k) > 0$ and (ii) $H'(k) > 0$. Let k_q and k_H be the unique solutions, if they exist, to $q(k) = 0$ and $H(k) = 0$ respectively. Clearly $k_H < k_q$. Also:

$$k \gtreqless k_H \Leftrightarrow d[q(k)/p(k)]/dk \gtreqless 0 \Leftrightarrow dmv/dk \gtreqless 0$$

along an optimal program. Therefore, as $p'(k) < 0$, $k > k_H$ implies c and mv are complementary and vice-versa. Alternatively, c and m are complementary on the positive side of the $q = 0$ locus and substitutable on the negative side providing q is not so negative as to make $H < 0$. By way of interpretation the following comments are made.

First as $q'(k) > 0$ it is never optimal for v to switch sign more than once and, furthermore, if the economy is growing ($\dot{k} > 0$) the switch, if it occurs, will be from $v = -1$ to $v = +1$ and vice-versa. Notice that if $m > 0$ in the asymtotic state then $v = +1$ if and only if $k(\infty) > k^*$, i.e. if and only if the economy is overdeveloped.

If the controls 'c' and 'mv' switch from being substitutes to complements it is necessary that $\dot{k} > 0$ and vice-versa. Also such a switch can occur at most once and requires $q < 0$ and $v = -1$. Furthermore, $k_H < k_q$ so this switch, if it occurs, precedes the switch in v if $k' > 0$ and vice versa.

If $c_o = 0$ then $q(k) > 0$ for all k (as $q = 0$ no longer exist in the region $k < k^*$). In fact, there exists a unique 'c_c' $\leq f(k^*) - k^*f'(k^*)$ such that $c_o < c_c$ implies $q(k) > 0$ for all k in which case, regardless of $k(0)$, $v = +1$ always. In other words if additional people are given weighting in the criterion function even for very low levels of per-capita consumption, it is always optimal to try to increase the rate of growth of the population. But if $c_c \leq c_o \leq f(k^*) - k^*f'(k^*)$ then this result is applicable only if $k(0)$ is not too small; otherwise per-capita income is so small that there will exist an initial segment of the optimal path that corresponds to $v = -1$, i.e., to reducing $\dot{L}/L$. Also, regardless of c_o, if $k(0)$ is sufficiently large and the asymptotic state is in an overdeveloped region then $v = +1$ always.

If $0 \leq c_o \leq f(k^*) - k^*f'(k^*)$ then a necessary condition for v to switch from -1 to +1 is that the economy pass from a state of lesser-development to over-development in which case the switch occurs while the economy is still in a lesser-developed state and vice-versa.

A particular example of these results is given by the following illustration. Suppose:

$$0 < c_o < f(k^*) - k^*f'(k^*)$$

and

$$k_H \text{ and } k_q \text{ exist.}$$

Assume $k(0)$ is sufficiently small so that $k(0) < k_H$ and therefore:

$$k(0) < k_H < k_q < k^* < k_1 .$$

Then the optimal path will asymptotically approach the k_1 stationary point where $k_1 > k^*$, i.e., the economy is overdeveloped and $u(c(k_1)) > 0$. Initially $v = -1$, $u(c(k)) < 0$ and also 'c' and 'm' are complementary so, as 'k' increases, both 'c' and 'm' increase until $k = k_H$. For $k > k_H$, 'c' increases and 'm' decreases to zero at $k = k_q$ at which point 'v' switches sign to +1. From then on both 'c' and 'm' increase together.

It is of interest to consider the manner in which the optimal programme changes if the assumptions placed on the population growth rate function $r(\cdot)$ are reversed, i.e.

$$r'(k) < 0 < r''(k) \qquad \text{for all } k.$$

It is again assumed that the size of the non-convexities is small. The only essential analytical change is a reflection of the $\Phi = 0$ locus about the vertical line $k = k^*$. Therefore most of the qualitative characteristics of the optimal programme (see Figures 7 and 8) will remain unchanged and so the discussion is restricted to results that stand in marked contrast to those applicable when $r'(k) > 0$. The Figures may be interpreted in the same manner as earlier.

If $0 \leq c_o \leq f(k^*) - k^*f'(k^*)$ there is not a threshold theory of growth for there is a unique Pontryagin stationary programme. This programme is characterized by $c(k(\infty)) > c_o$ and $q(k(\infty)) > 0$. In addition the asymtotic state is one of lesser-development, i.e. $k < k^*$. If $c_o > f(k^*) - k^*f'(k^*)$ either these results are still valid or all Pontryagin stationary programmes

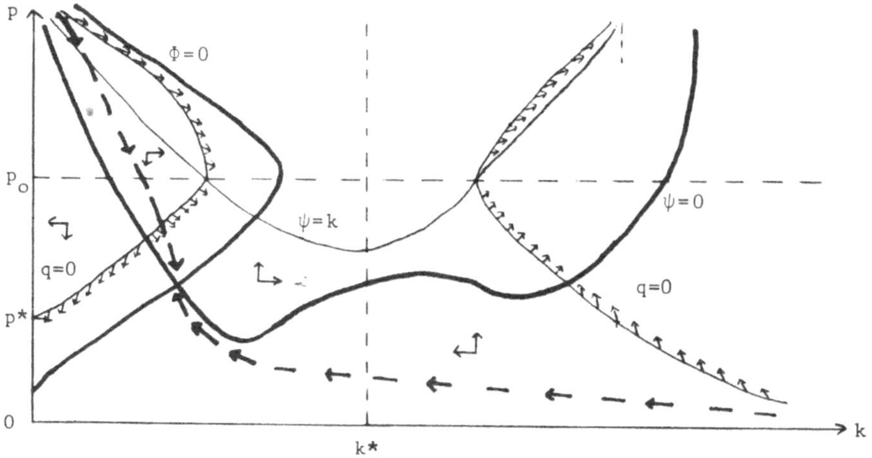

Figure 7: $c_o < f(k^*) - k^*f'(k^*)$

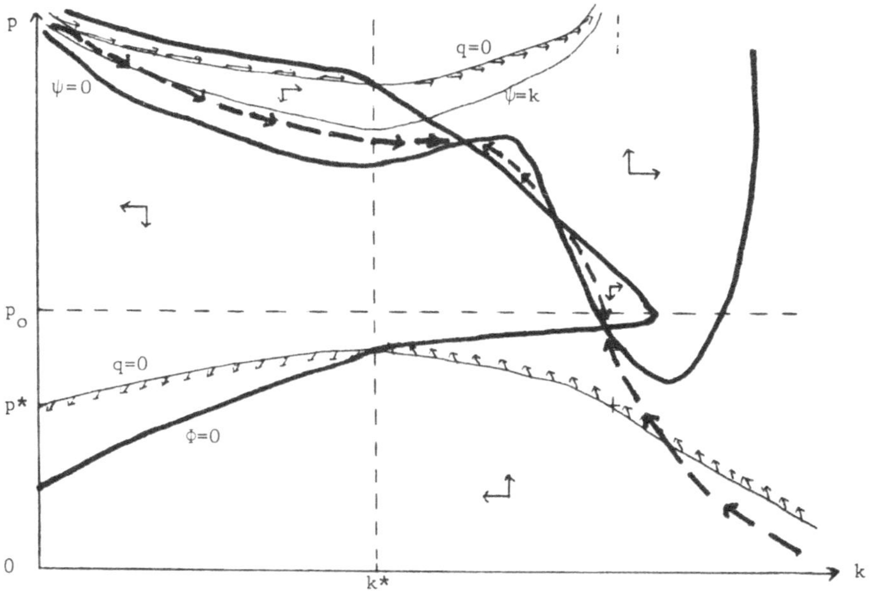

Figure 8: $c_o > f(k^*) - k^*f'(k^*)$

locate in the interval (k^*,∞) and there is still a threshold theory of growth. The scarcity price q of a unit of labor will be negative at all these points and also $c(\infty)$ will be less that c^* (as in Dasgupta's analysis). Per-capita utility will be negative at all Pontryagin stationary points except, possibly, at the one corresponding to the highest capital labor ratio.

(2) A Fixed Co-efficient Technology.

Now suppose that:

$$Q = \min\{\alpha K, \beta L\}$$

where α and β are the constant output-capital and output-labor ratios respectively. Then:

$$Q/L = \min\{\alpha k, \beta\} \equiv f(k), \text{ say.}$$

Notice that there is a discontinuity in $f'(k)$ if and only if $k = \beta/\alpha$. Also there is a surplus of labor (in terms of this technology) if $k < \beta/\alpha$ and surplus capital if $k > \beta/\alpha$. The production function is illustrated in Figure 9.

If it is assumed that $\dot{L}/L = n[f(k)]$, $n'(\cdot) > 0$ and $n''(\cdot) < 0$, then $r'(k) > 0 > r''(k)$ if $k < \beta/\alpha$ but $r(k)$ is constant otherwise. It will be assumed here that $r'(k) > 0 > r''(k)$ for all k; the results implied by the former assumption being easily inferred as a special case of those presented here.

Suppose first that $\rho \geq \alpha$. Then the $q = 0$ and $\Phi = 0$ loci can be shown to take the form attributed to them in Figures 10 and 11. Notice that:

$$q \equiv H - pk$$

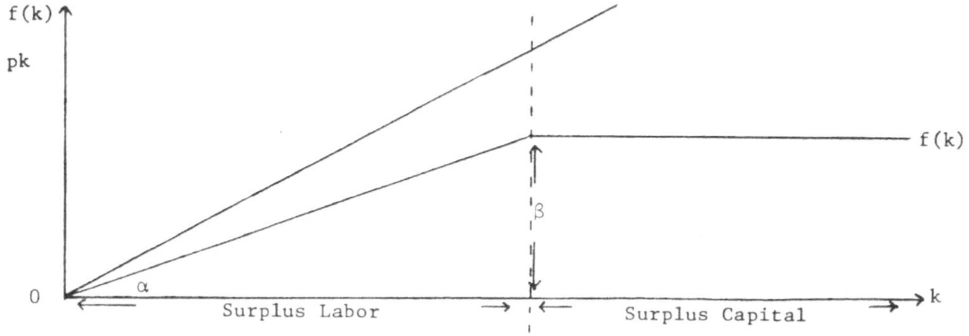

Figure 9.

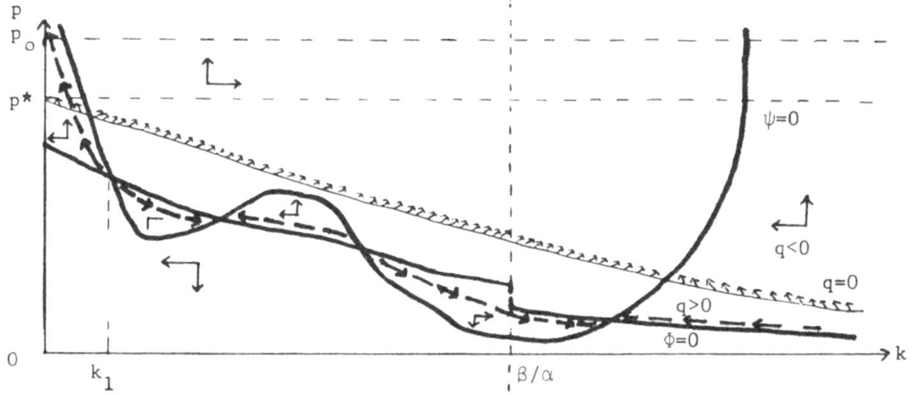

Figure 10. $\rho > \alpha$

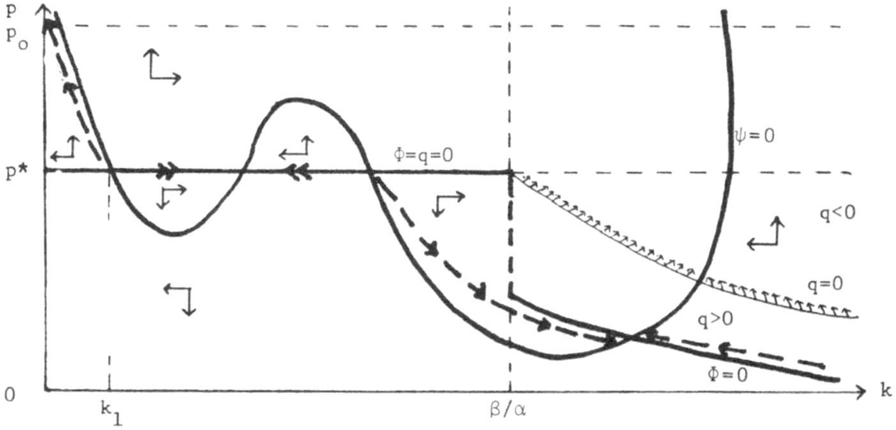

Figure 11. $\rho = \alpha$

so,

$$q[\rho - r(k)] = [u(c) - pc] + [f(k) - \rho k]$$

By assumption $f(k) - \rho k \leq 0$ so $q = 0$ requires $u(c) - pc \geq 0$ i.e. $p \leq p^*$. Furthermore the $\Phi = 0$ locus will be discontinuous at $k = \beta/\alpha$ as illustrated. This argument indicates why the $\Phi = 0$ and $q = 0$ loci do not take the form exhibited in the previous section.

In general the $\psi = 0$ locus may not exist. Let us suppose that the maximal sustainable level of per-capita consumption is no less than c^* i.e.

$$c_G \equiv \max_k [f(k) - kr(k)] \geq c^* .$$

This is a sufficient condition for the existence of the $\psi = 0$ locus; it is also necessary for the existence of Pontryagin stationary programs. Notice that this condition is the more likely to be violated according as $r(k)$ is higher for given k.

If this condition is violated, or more generally c_G is not so large that Pontryagin stationary programs exist, then the optimal program is the unique decaying turnpike (defined for all $k(0)$) with the property:

$$\lim_{\Delta \to \infty} p(\Delta) = \infty, \lim_{\Delta \to \infty} k(\Delta) = 0$$

Furthermore, along this program, $c(\Delta)$ and $k(\Delta)$ will be strictly monotonically decreasing functions of Δ.

On the other hand if c_G is sufficiently large that there do exist Pontryagin stationary programs the optimal program will be as illustrated by the dashed lines in Figures 10 and 11. If degenerate stationary programs are excluded then there are an even number of stationary programs with the particular property that the k_1 stationary program is locally unstable. If $k(0) < k_1$ then the optimal programme is the unique decaying turnpike described above. Furthermore there exists $t_o \geq 0$ such that $q < 0$ for all $t \geq t_o$. Also, if $k(0) \geq k$, then the usual results apply but, in particular, along the optimal program:

$$c \geq c^* > c_o \quad \text{and} \quad q(k) \geq 0$$

where equality implies $\rho = \alpha = f'(k)$. Notice that the non-negativity of 'q' is consistent with a surplus of labor in production.

The existence of the decaying turnpike is the most important difference between these results and those given earlier. It is therefore of interest to consider why an optimal program can take this form. Restrict attention to Pontryagin programs. Amongst such programs the optimal program $p = p(k)$ gives the maximum $c(\Delta)$, at any given Δ, consistent with the program not becoming infeasible in finite time. Now consider why this programme should decay ($\dot{k} < 0$) rather than grow for sufficiently low $k(0)$ i.e. all $k(0) < k_1$. Clearly:

$$p'(k) \leq 0 \text{ so } \dot{k} \geq 0 \Rightarrow \dot{p} \leq 0$$

i.e. only if $\dot{p} \leq 0$ can capital not be run down relative to the population. Recall that:

$$-\dot{p}/p = (\alpha - \rho) + r'(k)q/p.$$

Therefore, for such un-productive economies, i.e. those for which the 'net' marginal product of capital '$\alpha - \rho$' is negative (even for very low k which is in contrast to the case of a neo-classical technology) the return on capital is not sufficient to justify holding it unless the indirect benefit (q/p) is sufficiently positive. As $q/p \equiv [f(k) - \rho k] + [u(c) - cu'(c)]/p$ and $f(k) - \rho k \leq 0$ this requirement implies $c > c^*$, i.e. $p < p^*$. This cannot be so if $\rho = \alpha$ (see Figure 11) and even if $\rho > \alpha$ it is violated if $k(0)$ is sufficiently small (see Figure 10); for $p \geq p(k)$, otherwise the program is infeasible in finite time, and $p'(k) < 0$ together with $\lim_{\Delta \to \infty} p(k(\Delta)) = \infty$. Therefore $\dot{k} < 0$ for sufficiently small $k(0)$ and so must be negative for all $k(0) < k_1$.

Under the assumption $\rho \geq \alpha$ it can be verified that the size of the non-convexities is 'small' in the sense that $g \geq 0$. The existence of a Pontryagin program $p = p(k)$, which passes through all the stationary programmes and has the property $p'(k) \leq 0$ for all k, depends critically on this feature. Suppose now that $\alpha > \rho$. Then $g > 0$ if and only if $k > \beta/\alpha$ so the (non) convex region is equivalent to the region of surplus capital (labor) i.e. $\partial Q/\partial K (\partial Q/\partial L) = 0$, which in turn is equivalent to the over (lesser)-developed region.

As $\alpha > \rho$ it is now possible for $f(k) - \rho k$ to be positive so the $q = 0$ locus is as described in the previous section. In fact if c_o is sufficiently small (and certainly no greater than $(\beta/\alpha)(\alpha-\rho)$ then also the $\Phi = 0$ locus is an described in Figures 3 and 5 and the results of the analysis are as before. But for larger values of c_o the $\Phi = 0$ takes the form attributed to it in Figures 12 and 13. In particular it exists in the region delineated by $k < \beta/\alpha$ and $q < 0$ even when $c_o \leq (\beta/\alpha)(\alpha-\rho)$. The change in the form of the $\Phi = 0$ locus is a consequence of the size of the non-convexities being large.

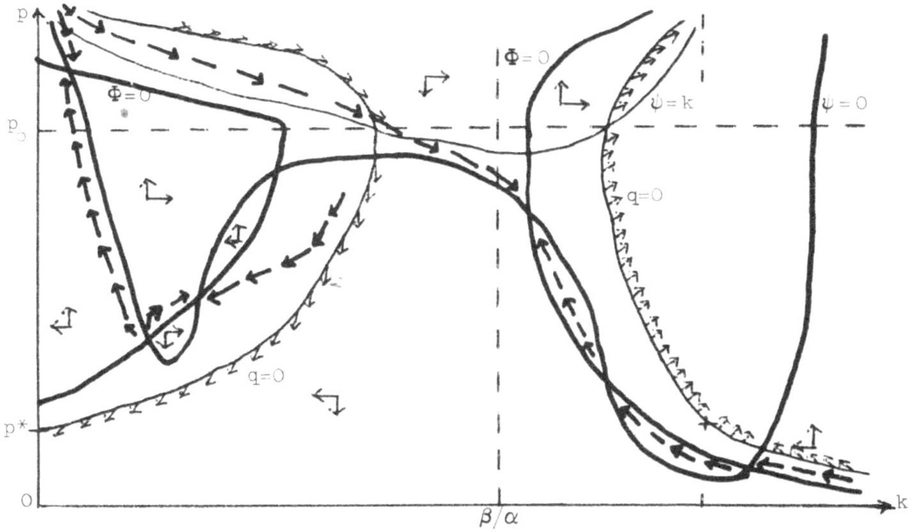

Figure 12. $0 < c_o < (\beta/\alpha)(\alpha-\rho)$

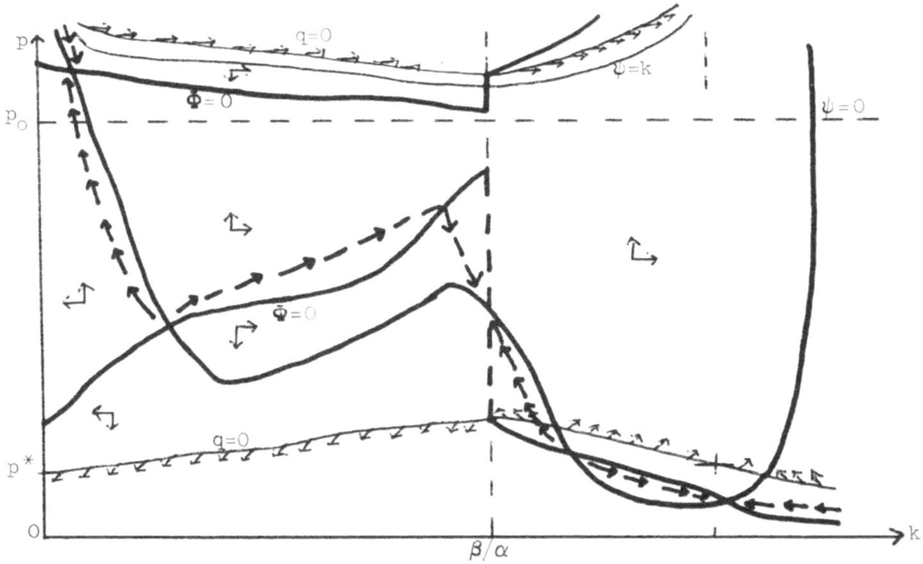

Figure 13. $c_0 > (\beta/\alpha)(\alpha-\rho)$

It is still true that there will be an uneven number of stationary programs, alternatively saddle points and totally unstable points, and the optimal program must approach such a point. But there may now be more than one such program (in contrast to the situation depicted in Figures 3 to 6 inclusive) and it is still an open question as to how one should choose between such programs. In certain cases it is possible to describe the optimal program but a complete solution is not available. For example if $k(0) \geq k_s$ (where $k_s \geq \beta/\alpha$ is the smallest capital labor ratio corresponding to a stationary program in the over-developed region) then it is never optimal to transfer into the lesser-developed region unless, possibly, if c_o is sufficiently large and certainly greater than $(\beta/\alpha)(\rho-\alpha)$. Also, under these circumstances, the optimal program is as described in the previous section.

Lastly, notice that p(k) may no longer be a monotonic decreasing function of k because the optimal program may be one that terminates in the non-convex region. Along such programs, and in contrast to those in the convex region ($k \geq \beta/\alpha$), there is a trade-off between 'c' and 'r'. Furthermore there appears little reason to think that c and/or $u(c)/(\rho-r)$, when evaluated at the Pontryagin stationary points, is an increasing function of k_i.

(3) The Dependency Ratio.

Between different countries there exists wide differentials in fertility rates. Restricting consideration to two countries with the same overall population growth rate the one with the higher fertility rate and lower infant mortality rate will have the higher dependency ratio [26]: that is, the ratio of the population measured in consuming units to the population measured in working units will be higher. The point is that the age structure, and hence the dependency ratio, is in large part determined by the fertility rate. A high dependency ratio has serious implications for the potential development of many countries; it will reduce private savings and the ability of the government to raise funds for development. In addition, a greater proportion of the savings and tax receipts will be absorbed by the needs of primary education, housing, and other social overheads. That this is quantitatively important is confirmed by a comparison of dependency ratios between developed and less developed nations. For instance, vis à vis the developed nations, many countries of Latin America and Asia have twice as high a percentage of their population under the age of fifteen as a result of the higher fertility rates; for these countries the absolute percentage is around forty-five to fifty.

The purpose here is to incorporate such considerations into our model. Define the dependency ratio as the inverse of:

$$D = L_w/L$$

where L is the size of the population and L_w the size of the work force. Implicit in this definition is the assumption that all members of the work force are alike and that all members of the population have the same consumption needs; this assumption will be retained. Implicit in the previous sections in the assumption that $D \equiv 1$ and the effect of relaxing this restriction is now examined.

If the population growth rate changes then the age and sex composition of the population will also change, i.e., the dependency ratio changes so:

$$D = d(r), \text{ say.}$$

But $r = r(k)$ and therefore:

$$D = d(r) \equiv d[r(k)] \equiv D(k), \text{ say.}$$

It is reasonable to assume that:

$$0 < D < 1.$$

Now the change in the composition of the population occurs over a period of time so the relationship $D = d(r)$ is strictly applicable only when r has remained constant for some period of time. Therefore this relationship is applicable for a comparative statics analysis but not for a dynamic analysis. It would be better, for instance, to assume D to be a weighted average of past growth rates. While this difficulty will be ignored, however, it is hoped that the analysis will still yield some insight into the effects of a flexible dependency ratio.

Changes in the population growth rate are primarily determined by changes in the fertility and infant mortality rates; more specifically the difference between the two, i.e., the 'net' fertility rate. Suppose, then, that r increases (decreases) due to an increase in the net fertility rate. It can be shown [26] that if the number of children in the population is large relative to the number of people in retirement, and this is certainly the case for most lesser-developed nations, then the dependency ratio increases (decreases) i.e.,

$$d'(r) < 0.$$

Therefore:

$$r'(k) \gtreqless 0 \Rightarrow D'(k) \lesseqgtr 0.$$

The function D(k) is depicted in Figure 14 where the 'c' variant pertains to the case where $r'(k) > 0$ for all k and the 'b' variant pertains to the case where there exists a unique k_d such that $r'(k_d) = 0$ and, in addition, $r''(k_d) < 0$.

The introduction of the dependency ratio does not change the form of the welfare function. However:

$$Q = F(K,L_w) = L_w F(K/L_w,1) = L_w F(k/D,1)$$
$$\equiv L_w \, f(k/D)$$

so:

$$Q/L = Df(k/D) \equiv f_*(k), \text{ say}$$

as D is a function of k only. Therefore:

$$\dot{K} = F(K,L_w) - C \Rightarrow \dot{K}/DL = F(K/DL,1) - c/D$$
$$= f(k/D) - c/D$$

so:

$$\dot{k}/k = \dot{K}/K - \dot{L}/L = \frac{\dot{K}}{DL} \cdot \frac{DL}{K} - \frac{\dot{L}}{L}$$

$$= f(k/D)\, D/k - c/k - r(k)$$

which implies:

$$\overset{\circ}{k} = Df(k/D) - kr(k) - c$$

$$= f_*(k) - kr(k) - c \ .$$

Therefore to introduce the dependency ratio into the analysis one need only replace f(k) by $f_*(k)$. Clearly if f(k) and $f_*(k)$ have the same qualitative properties then the qualitative nature of our results will be invariant under this transformation. Whether or not this is so depends on the form of D(·). Notice that the transformed production function, i.e., $f_*(k)$, pertains only to steady state values of k; as mentioned earlier, this difficulty is being blatantly ignored.

The dependency ratio is likely to be much larger in the lesser-developed nations in which the demographic transition has occurred over a relatively short period of time. In addition, it is preceisely the lesser-developed nations that are more likely to be adequately characterized by a fixed coefficient technology. Therefore, in order to maximize the possibility that the transformation of f(k) to $f_*(k)$ (i.e., $D \equiv 1$ to $D = D(k)$) will lead to significant qualitative changes in our results, a fixed coefficient technology will be assumed in what follows. Suppose:

$$f(k) = \begin{cases} \alpha k & \text{if} \quad k \leq \beta/\alpha \\ \beta & \text{if} \quad k \geq \beta/\alpha \end{cases}$$

Then:

$$f_*(k) \equiv Df(k/D)$$
$$= \begin{cases} \alpha k & \text{if} \quad k \leq {}_*k \\ \beta D(k) & \text{if} \quad k \geq {}_*k \end{cases}$$

where:

$${}_{*}k/D({}_{*}k) = \beta/\alpha$$

Therefore $\quad {}_{*}k/(\beta/\alpha) = D({}_{*}k) < 1$ by assumption and ${}_{*}k$ now plays the role of β/α in the previous section. Therefore:

$${}_{*}k < \beta/\alpha \ .$$

Also if $k > {}_{*}k$ then:

$$f(k) > f_{*}(k) > 0 \text{ as } 0 < D < 1$$

$$f_{*}'(k) = \beta D'(k)$$

$$f_{*}''(k) = \beta D''(k) \ .$$

The functions $f(k)$ and $f_{*}(k)$ are illustrated in Figure 15. The 'oab' and 'oac' variants of $f_{*}(k)$ correspond, respectively, to the 'b' and 'c' variants of $D(k)$ in Figure 14. The 'oab' variant will also vary depending on the relationship between k_d and ${}_{*}k$. Notice, in particular, that depending on the sign of $D'(k)$ and $D''(k)$ it is possible to have a negative marginal product of capital (i.e., $f_{*}'(k) < 0$) and/or increasing returns to capital alone (i.e., $f_{*}''(k) > 0$). Consider the marginal product of capital. Clearly:

$$f_{*}'(k) = f'(k/D) + D'(k)[f(k/D) - (k/D)\ f'(k/D)]$$

i.e., an increase in capital per head (k) leads to an increase in capital per worker (k/D) by an even greater amount and, in turn, the effect of this on output per head may be decomposed additively into two parts, namely:

(i) The direct effect on output per head, via an increase in output per worker, i.e., $f'(k/D)$, and

(ii) The indirect effect on output per head via the induced change in L_w/L evaluated at the marginal product of one worker $[f(k/D) - (k/D)\ f'(k/D)]$, i.e., $D'(k)[f(k/D) - (k/D)f'(k/D)]$

If the sum of (i) and (ii) is negative, and this necessitates $D'(k) < 0$, then the marginal product of capital is negative. Suppose $D'(k) < 0$, that there is a fixed coefficient technology, and $k > {}_{*}k$. Then:

$$k/D(k) > \beta/\alpha \quad \text{so} \quad f'(k/D) = 0 \ .$$

Also:

$$D'(k)[f(k/D) - (k/D)\ f'(k/D)]$$

$$= D'(k)\ f(k/D) = \beta D'(k) \quad \text{as} \quad k/D(k) > \beta/\alpha$$

$$< 0 \quad \text{as} \quad D'(k) < 0 \ .$$

Therefore the direct effect (i) is zero and this is dominated by the indirect effect (ii) so the marginal product of capital is negative as illustrated by 'ac' in Figure 15.

Now let us consider whether or not the qualitative results of the previous section will change with this introduction of the dependency ratio. It is assumed directly that $r'(k) > 0 > r''(k)$ for all k so it is the 'oac' variant of $f^*(k)$ which is relevant.

If the $\psi = 0$ locus exists it will take the usual form if the extra condition:

$$\lim_{k \to \infty} D'(k) = \lim_{k \to \infty} d'(r) \lim_{k \to \infty} r'(k) = 0$$

is imposed; this condition guarantees that $f_*(k) - kr(k)$ is negative for all sufficiently large k. Also the $q = 0$ locus will take the usual form providing $k = \beta/\alpha$ is replaced by $_*k$. By analogy with our earlier results:

$$[0, {}_*k] \text{ is a convex (non-convex) region}$$
$$\text{if and only if } \rho \geq (<) \alpha$$

where a convex region requires $g \geq 0$ and vice-versa. Also, for sufficiently large k, the $q = 0$ locus must approach the k-axis and this constrains the $\Phi = 0$ locus to be downward sloping i.e., the interval $[k_c, \infty)$ is convex for sufficiently large k_c. However, and in contrast to our earlier results, there may now exist a non-convex region in the interval $({}_*k, k_c)$.

It has already been observed that a complete analytical solution to this problem is not presently possible if there exist non-convex regions. Therefore, rather than considering the general case and in an attempt to obtain some insight as to the effect of the dependency ratio on the optimal program, only the case $\rho = \alpha$ is illustrated (see Figures 16 and 17 which should be compared with Figure 10 - also, as it is not central to our purpose, the $q = 0$ locus is not illustrated but it can be superimposed from Figure 10). In Figures 16 and 17 the effect of introducing the dependency ratio is indicated by the transformation of β/α to $_*k$ and the

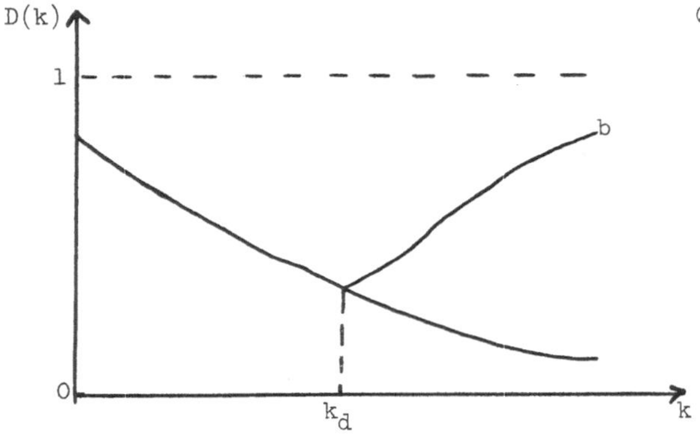

Figure 14.

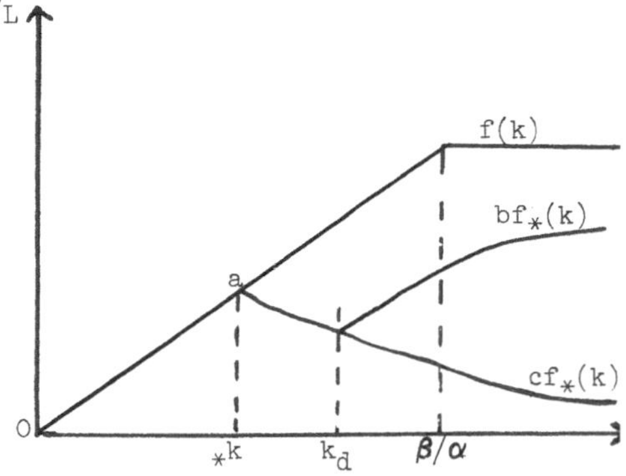

Figure 15.

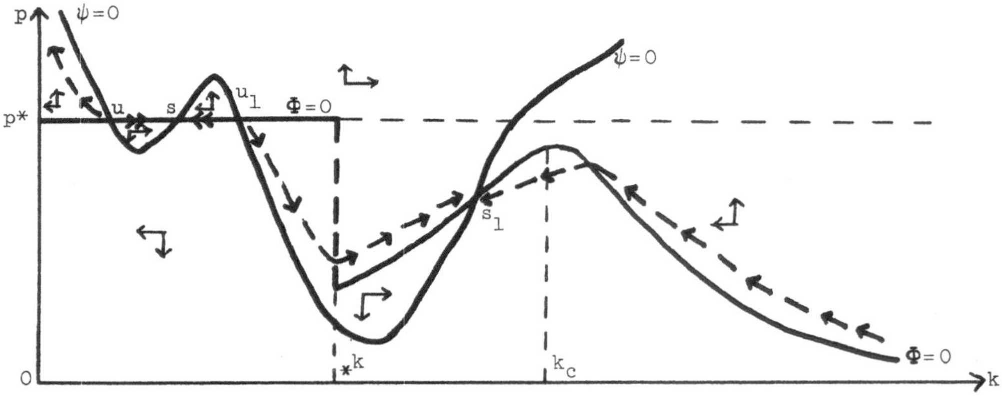

Figure 16. $\rho = \alpha$

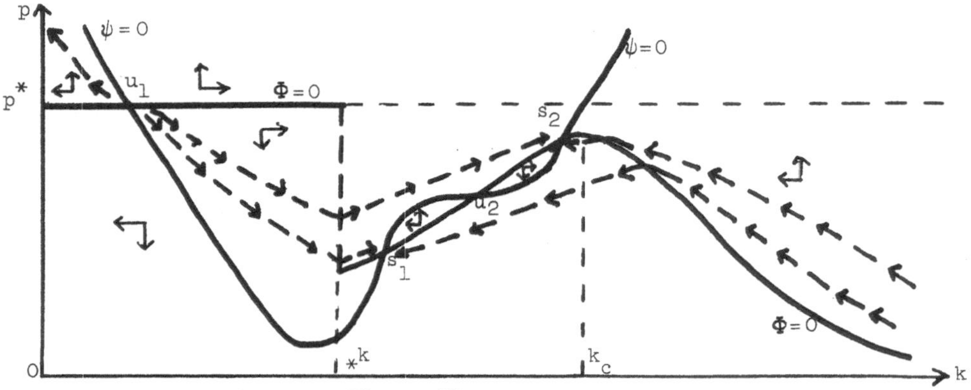

Figure 17. $\rho = \alpha$

existence of the non-convex region $({}_*k, k_c)$; in this region the $\Phi = 0$ locus must be upward sloping i.e. $dp/dk|_{\Phi=0} > 0$. In general there may exist multiple stationary points in this region; in Figure 16 a unique such point is shown and in Figure 17 three such points. Also, in total, there must exist an even number of stationary points alternating between totally unstable (u) and saddle points (s).

Because there must be a Pontryagin path passing through every point in the (p,k) phase space it is possible to show (i) whilst there may be a limit cycle around u_2 there cannot be one around u_1 and (ii) there must be a unique program emanating from u_1 that approaches s_1 (this will also apply to s_2 if there is not a limit cycle around u_2 ... but if there is such a limit cycle then this need not be the case).

In Figure 16 there is, therefore, a unique programme $p = p(k)$ passing through all stationary points; this is the optimal program In contrast to our earlier results

(a) for certain k(0)'s per-capita consumption is not a monotonic function of k and

(b) in the final phase of the optimal programme per-capita consumption is a decreasing function of k, i.e. $p'(k) > 0$.

However (see Figure 17), if there are multiple stationary points in the non-convex region, then there are two candidates for optimality, $p_1(k) \geq p_2(k)$ say. It is seen that, whichever is the optimal program, properties (a) and (b) above must still hold.

It is of interest that, because $p'(k)$ is no longer one-signed, it follows that $q'(k)$ is no longer one-signed. But, nevertheless, $q(k)$ can still only switch sign at most once. Therefore, if the population growth rate can be controlle directly, then our earlier results pertaining to whether it should be increased or decreased remain unchanged.

Due to the inherent non-convexity of this model the results are somewhat limited. It is hoped, however, that the formulation of a model which includes a flexible dependency ratio is of interest. Certainly such a model is amenable to simulation techniques and this is the subject of ongoing research. If the Pontryagin programmes do exhibit the configuration shown in Figure 17 then the analysis on p. 18 indicates $p = p_2(k)$ to be the optimal programme; for $p = p_1(k)$ cannot be the T-limit of finite horizon optimal programme. Such a program has the property that amongst all Pontryagin paths (and for each Δ) $c(\Lambda)$ is maximized subject to the constraint that this program not become infeasible in finite time. This is certainly a property of the optimal program when there are no non-convex intervals; and it is conjectured that it could form the foundation for a more general sufficiency theory than presently exists. Some would conjecture that it is never optimal to approach a Pontryagin stationary point in a non-convex interval; the results of this section indicate that this conjecture is quite definitely not valid.

5. EXHAUSTIBLE RESOURCES

In this section it is supposed that there exists a non-renewable resource in fixed and finite supply which can be extracted at zero cost. The existing literature on the rate at which such an exhaustible resource should be depleted generally focuses attention on the possible mitigating effects of capital accumulation and/or technological change in the growth process. Population is assumed exogenous to these models and so plays only a limited role [10, 18] . This assumption is now relaxed; our interest being the interdependence between alternative resource depletion policies and population policies. However, in order both to simplify the analysis and to illuminate better the role of population policy, total consumption is identified with the quantity of the resource extracted in each period and capital accumulation is ignored; a full synthesis incorporating population, capital and exhaustible resources will not be attempted here. Furthermore, only "first best" solutions will be considered.

(i) The model

Instantaneous per-capita utility is assumed to be given by:

$$w_t = w(C_t/L_t, X_t/L_t, L_t) = w(c_t, x_t, L_t) \text{ say}$$

where X_t is the stock of the resource remaining at time t. The second argument in the utility function cannot indicate that the stock of the resource has value as a source of future consumption; for this would imply double counting. However the stock may well be valued by society for its own sake. It may be considered to represent a wealth effect or a conservation motive depending on one's interpretation. This formulation allows for existing consumption to be valued less highly as the resource is depleted e.g. when the resource is associated with leisure. The 'x' term could also be a dummy variable reflecting society's uncertainty concerning the possibility of finding a substitute resource, from which consumption goods can be produced, when the present resource is exhausted. The third argument allows for a population externality, e.g. congestion factors, in individual preferences. To facilitate both the analysis and our understanding of the effect of this population externality on the optimal policy, the function 'w' is assumed to take the specific form:

$$w_t = L_t^{\alpha-1} u(c_t, x_t).$$

Now suppose it is desired to choose the per-capita consumption profile (c_t), the terminal time T and the population growth rate n_t say, so as to maximise the classical utilitarian social welfare function:

$$(1) \qquad W_T = \int_0^T e^{-\rho t} L_t \, w_t \, dt \qquad \rho \geq 0$$

$$= \int_0^T e^{-\rho t} L_t^{\alpha} \, u(c_t, x_t) dt$$

subject to

$$(2) \qquad \int_0^T C_t dt \leq X_o > 0 \qquad C_t \geq 0$$

$$(3) \qquad \underline{n} \leq n_t \leq \bar{n}, \quad t \epsilon [0,T)$$

(4) X_o and L_o given, $\underline{n}$ and $\bar{n}$ finite and given

where X_o is the initial stock of the resource. The parameter α will be allowed to take any value in the interval $[0, 1]$; notice that as α decreases the individualistic population externality increases in importance and, when $\alpha = 0$, it is exactly counterbalanced by the "adding up" effect induced by representing social welfare as the sum of per-capita welfares.
In effect, by varying α in the above range, one can consider the alternative

implications of both average and classical utilitarianism and also the intermediate cases.

This social welfare function is not invariant to linear increasing transformations of the utility index, as discussed earlier, even in the average utilitarian case $(\alpha = 0)$ for T is now endogenous. It is assumed that $u(c,\dot{x}) = 0$ implies a unique solution $c = c_o(x)$ say and that $c_o(x) > 0$. It is also assumed that $u(c,x) - cu_c(c,x) = 0$ implies a unique solution $c = c^*(x)$. Clearly $c^*(x) > c_o(x)$ for all x.

Let the amount of the resource left over at time t be:

$$X_t = X_o - \int_o^t C_t \, dt.$$

Then Eq.(2) is equivalent to:

(5) $$dX_t/dt = -C_t \ , \ X_t \geq 0 \qquad X_o > 0 \ .$$

As $x_t = X_t/L_t$ Eq.(5) can be re-written:

(6) $$dx_t/dt = -\left[c_t + n_t x_t\right] \qquad x_T \geq 0 \qquad x_o > 0$$

and L_t is non-zero and finite on the interval $[0, T]$. Notice also that $X_T \geq 0 \Rightarrow X_t \geq 0$ for all $t \Rightarrow x_t \geq 0$ for all t.

Now the model is restricted by assuming that:

(7) $$\rho - \alpha\bar{n} \geq \delta > 0 \quad \text{and} \quad \bar{n} \geq 0$$

Define:

(8) $$\Delta(t) = \rho t - \alpha\left[\int_o^t n_s \, ds\right]$$

which is a strictly increasing transformation of t and ranges from zero to infinity when t ranges from zero to infinity. Therefore the model given by Eqs. (1), (3), (4) and (6) reduces to the choice of a per-capita consumption profile (c), a population growth rate profile (n) and the terminal time $\Delta(T)$ so as to maximise:

(9) $$W_{\Delta(T)} = \int_{o}^{\Delta(T)} e^{-\Delta} u(c,x)/(\rho - \alpha n).d\Delta$$

subject to:

(10) $$dx/d\Delta = -[c + xn]/(\rho - \alpha n)$$

and the feasibility constraints;

(11) $\underline{n} \leq n \leq \overline{n}$, x_o, $\underline{n}$ and $\overline{n}$ finite and given, $c \geq 0$ and $x \geq 0$,

where all variables are implicitly functions of Δ.

The programme optimal relative to a finite and arbitrary horizon $\Delta(T)$ is determined first; it will be called a T-optimal programme and denoted $c = c^T(x)$. The infinite horizon optimal programme is denoted by $c = c^{\infty}(x)$. The second step is to determine the optimal $\Delta(T)$.

The current value Hamiltonian is:

(12) $$H = u(c, x)/(\rho - \alpha n) - p[c + xn]/(\rho - \alpha n) + \mu_1(\overline{n} - n) + \mu_2(n - \underline{n}) + rx$$

where μ_1, μ_2 and r are the Lagrange multipliers and p is the continuous co-state variable, i.e. the scarcity value of the resource per-capita, which corresponds to 'x'.

Assume that utility is increasing and strictly concave in both arguments; and also that:

$$u_c(0,x) = \infty \quad \text{and} \quad u_c(\infty, x) = 0 \text{ for all } x$$

where u_c is the marginal utility of consumption.

Define:

(13) $$q(\alpha) \equiv \alpha H - px$$

Clearly:

(14) $\quad \mu_1(\bar{n} - n) = \mu_2(n - \underline{n}) = rx = 0$ and $\mu_1, \mu_2, r \geq 0$

and the controls 'c' and 'n' will indeed be maximising if chosen to satisfy:

(15) $\quad u_c(c,x) = p > 0$ and $q = \mu_1 - \mu_2$.

It follows that;

(16) $\quad c = c(p, x)$ say, $\partial c/\partial p < 0$ and $c(p, x) > 0$

for all x and for all finite p. Also:

(17) $$n = \begin{cases} \bar{n} \\ \underline{n} \end{cases} \leftrightarrow q(\alpha) \gtrless 0$$

so $q(\alpha) = 0$ is the switching locus on which n may take an interior value.

Now $H_x = (u_x - pn)/(\rho - \alpha n)$ when $x > 0$ and therefore p must satisfy the differential equation:

(18) $$dp/d\Delta = p - H_x$$
$$= \{p\,[\rho + n(1 - \alpha)] - u_x\}/(\rho - \alpha n)$$

Pontryagin programmes, henceforth labelled Pn-programmes, are those that satisfy the above necessary conditions for optimality i.e. Eqs. (10) and (16) to (18) inclusive.

Furthermore $H_{xx} = u_{xx} \leq 0$ so the transversality conditions

(19) $$\lim_{\Delta \to \Delta(T)} e^{-\Delta} p(\Delta) \geq 0 \qquad \lim_{\Delta \to \Delta(T)} e^{-\Delta} p(\Delta) x(\Delta) = 0$$

are necessary and sufficient for optimality if T is finite; and sufficient if T is infinite. If $\lim_{\Delta \to \Delta(T)} e^{-\Delta} p(\Delta) > 0$, as is typically the case, then Eq.(19) reduces to the efficiency requirement:

$$(20) \quad x(\Delta(T)) = 0$$

i.e. along a finite horizon optimal programme the resource must be completely exhausted at the terminal date.

The term $q(\alpha)$ has the same interpretation and plays the same role as its counterpart in Section 4. In fact, if q_L is the scarcity value of a member of society, the analysis of this model in extensive form indicates that:

$$dz/d\Delta = z - \left[\alpha u - cu_c - xu_x\right]/(\rho - \alpha n)$$

where $z = L^{1-\alpha} q_L$. Assume (i) $q_L(\Delta(T)) = 0$ and (ii) $dn/d\Delta = 0$ for all Δ. Then one can verify that the solution to this equation is:

$$z \equiv L^{1-\alpha} q_L = q(\alpha)\left[1 - e^{-(\Delta(T)-\Delta)}\right]$$

Along Pn-programmes that approach a Pn-stationary point, i.e. $dx/d\Delta = dp/d\Delta = 0$, and for which (ii) holds, the solution is:

$$z \equiv L^{1-\alpha} q_L = q(\alpha)$$

As q_L must be continuous, and because it will be seen below that Pn-programmes cannot violate condition (ii) on an interval of Δ, condition (ii) can be dropped. It follows that:

$$q_L \gtreqless 0 \leftrightarrow q(\alpha) \gtreqless 0 \quad \underline{\text{or}} \quad \Delta = \Delta(T)$$

In particular this scarcity value is zero only on the switching locus $q(\alpha) = 0$ and at the terminal date $\Delta(T)$. Observe that for any finite T,

$L(\Delta(T)) \neq 0$, so the transversality conditions require q_L to be zero on any T-optimal programme at the terminal date; as is indicated by the above relationship between q_L and $q(\alpha)$.

The second step in the analysis is the determination of an optimal T in the interval $[0, \infty]$. It will be denoted by $\hat{T}(\alpha)$. A result in control theory (see [43] for example) indicates that if an interior optimum exists it satisfies:

$$\lim_{\Delta \to \Delta(T^*)} e^{-\Delta} H(\Delta) = 0 \tag{21}$$

i.e. for some finite T^*,

$$c(\Delta(T^*)) = c^*(x(\Delta(T^*)))$$

$$= c^*(0) \tag{22}$$

for it will be seen below that $x(\Delta(T)) = 0$ for all T-optimal paths. Setting $T = 0$ is trivially non-optimal so either $c(\Delta(T^*)) = c^*(0)$ or $T = \infty$.

Eq. (21) may have multiple solutions, $T^*_i(\alpha)$ say, some of which correspond to local minima; or the solution may represent a unique minimum so $\hat{T}(\alpha) = \infty$. An additional result will therefore be needed to provide a ranking of T-optimal programmes.

It is possible to formulate the model in extensive form and, even without the assumption $\rho - \alpha\bar{n} \geq \delta > 0$, show that the necessary conditions can be manipulated to yield those given above; the analysis and the phase diagrams below remain valid in the region $q(\alpha) \leq 0$; and in the region $q(\alpha) > 0$ if $\rho + \bar{n}(1 - \alpha) > 0$. This is not the case in Section 4 where the population growth rate is endogenous. This assumption serves to keep the presentation in a unified form. It also indicates that the analysis in Section 4 can be subjected to the same type of sensitivity analysis, with respect to the parameter α, as in the next sub-section.

Before analysing the necessary conditions it is useful to define certain functions. Their properties are summarised here, without proof, and will be utilised in the sub-sections below.

If $q(\alpha) \equiv q(c, x; \alpha) \equiv \alpha H - px = 0$, where $p = u_c(c, x)$ then:

(i) $\quad q_c = -u_{cc}(\alpha c + \rho x)/(\rho - \alpha n)$

(ii) $\quad q_x = [(\alpha u_x - \rho u_c) - (\alpha c + \rho x)u_{cx}]/(\rho - \alpha n)$

(23) (iii) $\quad dq(\alpha)/d\Delta = x(u_x - \rho u_c) + cu_c(1 - \alpha)$

(iv) $\quad \partial q(\alpha)/\partial\alpha > 0$

(v) $\quad c \geq c^*(x) > c_o(x)$ with equality $\leftrightarrow x = 0$

Also the locus $q(\alpha) = 0$ is independent of 'n'.

Define $\Theta^\alpha(c, x) = \alpha u_x - \rho u_c$. Then:

(24) (i) $\quad \Theta^\alpha_c = \alpha u_{xc} - \rho u_{cc}$
(ii) $\quad \Theta^\alpha_x = \alpha u_{xx} - \rho u_{cx}$
$\Longrightarrow$ $dc/dx > 0$ if $u_{cx} \geq 0$ and $\Theta^\alpha(c, x) = 0$.

Define $\Phi^\alpha(c, x) = dc/d\Delta$. Then:

(i) $\quad \alpha = 1 \Rightarrow dc/d\Delta = \Phi^1(c, x) \gtreqless 0 \leftrightarrow \Theta^1(c, x) \gtreqless 0$

(ii) $\quad$ if $\Phi^1(c, x) = 0$ and $\alpha < 1$ then $q(\alpha) \gtrless 0 \Leftrightarrow \Phi^\alpha(c, x) \lessgtr 0$

(25) (iii) $\quad \Theta^\alpha \geq 0 \Rightarrow \Phi^\alpha(c, x) \geq 0$

(iv) $\quad \Phi^\alpha_{x.}(\rho - \alpha n) = [\rho + n(1 - \alpha)]\, u_{cc} - u_{xc}$
$\quad \Phi^\alpha_{c.}(\rho - \alpha n) = [\rho + n(1 - \alpha)]\, u_{cc} - u_{xc}$

(v) $\quad$ if $u_{cx} \geq 0$ and $\rho + n(1 - \alpha) > 0$ then $\Phi^\alpha_x > 0$ and $\Phi^\alpha_c < 0$.

In the diagrams below arrows are used to indicate the direction of movement of c and x as Δ increases. A '+' on one side of the $\theta^{\alpha} = 0$ or $q(\alpha) = 0$ loci indicates the positive side of these loci and vice-versa.

The locus of points for which $dx/d\Delta = 0$ is illustrated for the case $\underline{n} < 0$. This locus must include the origin in (c, x) space and, as x increases from zero, it will first have a constant slope $-\underline{n}$ in the $q(\alpha) < 0$ region and, if it eventually meets the $q(\alpha) = 0$ locus, it will coincide with it for a unique choice of $n \in [\underline{n}, \bar{n}]$ and $n < 0$.

On the other hand:

(i) $\underline{n} = 0$; $dx/d\Delta = 0 \leftrightarrow c = 0$, x arbitrary.

(ii) $\underline{n} > 0$; $dx/d\Delta = 0 \leftrightarrow c = x = 0$.

With this information we can now turn to the analysis of the optimal policy.

(ii) A Population Externality

A useful starting point in the study of optimal growth theory in the presence of an exhaustible resource is provided by Gale's "optimising cake-eater" example [12]. The sole member of society, who lives for a finite length of time, T say, and does not discount against future utilities ($\rho = 0$), can consume a slice C_t in each period of a non-perishable cake of finite size X_o. Assuming that his preferences can be represented by the intertemporal sum of his instantaneous utilities, i.e. $\int_o^T u(C_t)dt$, then the optimal programme must satisfy:

$$c_t \equiv C_t = X_o/T \text{ for all } t$$

if $u(\cdot)$ is increasing and strictly concave. But then as $T\to\infty$ the only feasible constant consumption programme is given by $C_t = 0$, for all t, which is clearly non-optimal, i.e. an optimal programme does not exist. Positive discounting of future utilities in the infinite horizon case is neccessary for the existence of an optimal solution and is sufficient given suitable behaviour of

$u(c)$ as $c \to 0$ [12].

Koopmans [20, 21] has modified this model in three directions:

(a) T is determined optimally;

(b) there is discounting of future utilities at the constant non-negative rate ρ;

(c) there is a subsistence level of per-capita consumption $\underline{c}$, assumed less than c^* which is the unique solution to $u(c) - cu'(c)$, below which all life ceases instantly.

The population size L is assumed given and constant for the first T periods but zero for all $t > T$. The maximum survival period is $X_0/L\underline{c}$, by (c), which is finite and so the non-existence problem referred to above does not arise. The essential point raised by Koopmans is that there is a trade-off between the optimal survival time and optimal consumption rates.

Observe that this model, if $\rho > 0$, is the special case of the model in subsection (i) obtained by setting $\alpha = 1$ (no individualistic population externality), $u_x(c, x) \equiv 0$ (no 'conservation' motive), $\underline{n} = \bar{n} = 0$ (no inter-temporal population policy) and $c_t \geq \underline{c}$. The Koopmans model captures some aspects of the population problem but now, rather than choosing the population profile from a set of step functions with the same maximum height, the feasible set of population policies has been enlarged to include all population profiles with growth rates between prescribed upper and lower bounds.

Throughout this section it is assumed that:

$$u = u(c)$$

so there is no wealth effect or conservation motive.

Therefore Eqs. (23) to (25) inclusive can be greatly simplified for $u_x = u_{cx} = 0$. Eq. (23) indicates that the locus $q(\alpha) = 0$ has positive slope in (c,x) space and emanates from the point $(c^*,0)$. Above this locus $q(\alpha) > 0$ and vice-versa.

From Eq. (25) notice that if $\rho + \underline{n}(1-\alpha) > 0$ then $dc/d\Delta < 0$ everywhere; the same holds true for all other parameter configurations providing $q(\alpha) > 0$. If $\rho + \underline{n}(1-\alpha) < 0$ then $q(\alpha) < 0$ implies $dc/d\Delta > 0$ and the locus $q = 0$ separates the regions in which $dc/d\Delta > 0$ and < 0 respectively.

Now part (iii) of Eq. (23) implies:

(26) $$dq(\alpha)/d\Delta = (1-\alpha)c - \rho x\, u'(c).$$

If $dq(\alpha)/d\Delta = 0$ then $c = \rho x/(1-\alpha)$. Let this ray meet $q(\alpha) = 0$ at the point $(\tilde{c},\tilde{x})$. Then:

$$\tilde{c} \text{ solves } \alpha u(c) - cu'(c) = 0$$

Therefore Pn-programmes cross from the negative to the positive side of the $q(\alpha) = 0$ locus if $c < \tilde{c}$ $(x<\tilde{x})$ and vice-versa.

Suppose $\alpha = 1$ then $(\tilde{c},\tilde{x}) = (c^*,0)$. Now let α decrease. The $q(\alpha) = 0$ locus still emanates from $(c^*,0)$ but becomes steeper. The ray $c = \rho x/(1-\alpha)$ becomes flatter. Therefore $(\tilde{c},\tilde{x})$ continually increases i.e. moves upwards and to the right in (c,x) space. If $\rho + \underline{n}(1-\alpha) > 0$ then $dx/d\Delta$ is negative at this point. If equality holds then $dx/d\Delta = 0$ and $\tilde{c} = -\underline{n}\tilde{x}$. Now suppose $\rho + \underline{n}(1-\alpha) < 0$. Then the choice of 'n' which yields $dx/d\Delta = 0$ at $(\tilde{c},\tilde{x})$ also yields $dc/d\Delta = 0$. Therefore $(\tilde{c},\tilde{x})$ is a Pn-stationary point which is approached in finite time. No other points along $q(\alpha) = 0$ can have this property. Furthermore $(\tilde{c},\tilde{x})$ is a saddle-point.

With the above information one can construct the phase diagrams; see Figures (18) to (20) inclusive which correspond to different parameter configurations and in which, implicitly, $\underline{n} < 0 \leq \bar{n}$.

For the remainder of this section the classical utilitarian case, $\alpha=1$, will be examined in depth; the interpretation of the other cases can then easily be read off from the Figures. In the next section the horizon will be determined endogenously and then this "full" optimal programme will be discussed for all parameter configurations; these programmes are indicated by continuous arrowed lines.

If $\alpha=1$ then Figure 18 is applicable but with a small modification; the point $(\tilde{c},\tilde{x})$ coincides with $(c^*,0)$ and the unique Pn-programme which approaches $(c^*,0)$ lies everywhere in the $q(1) > 0$ region.

As $\alpha=1$ Eq.(8) implies:

(27) $$dc/d\Delta = \rho u'(c)/u''(c)$$

which is equivalent to the more familiar form;

$$e^{-\rho t}\, u'(c_t) \text{ constant for } 0 \leq t \leq T$$

and so is recognised as the condition which guarantees that shifting a small amount of consumption from a neighborhood in $[0,T]$ of t' say, to one of t", cannot lead to an increase in social welfare.

Clearly per-capita consumption is a strictly decreasing function of "time" along all Pn-programmes and also:

(28) $$\begin{cases} c(\Delta) \to 0 \quad \leftrightarrow \quad \Delta \to \infty \\ p(\Delta(T)) > 0 \text{ for finite } T. \end{cases}$$

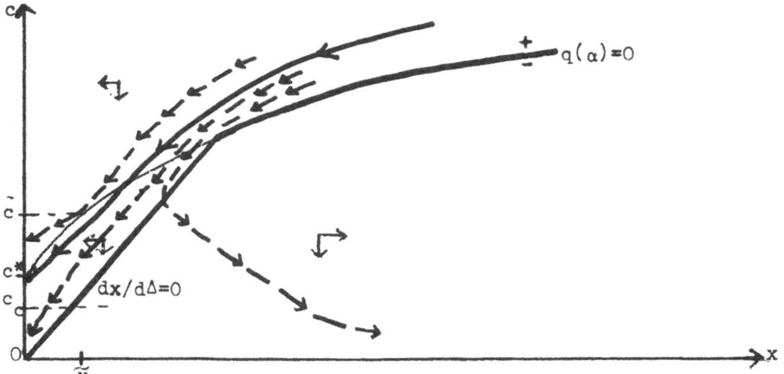

Figure 18: ρ+n(1-α) > 0

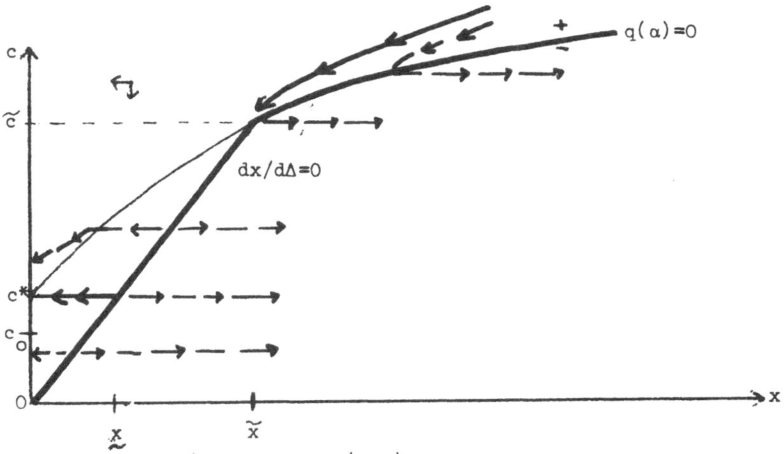

Figure 19: ρ+n(1-α) = 0

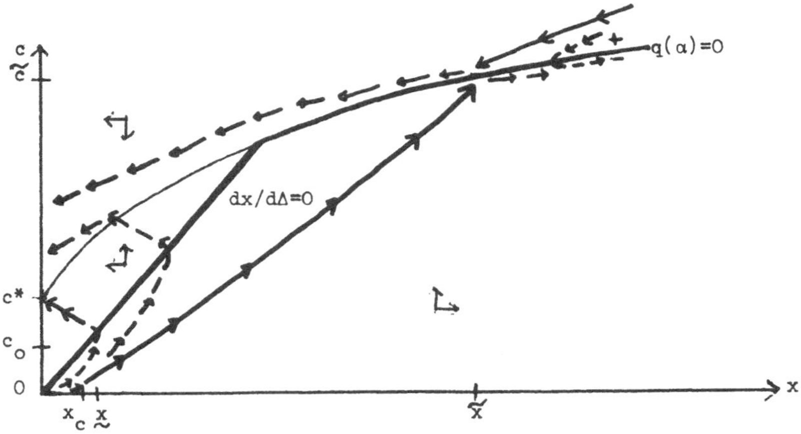

Figure 20: ρ+n(1-α) < 0

Consider the subset of Pn-programmes, which do not approach the origin and which are defined for all $\Delta \geq 0$ i.e. all Pn-programmes for which $c<c^{\infty}(x)$. If $\underline{n} > 0$, then $dx/d\Delta < 0$ everywhere, and there are no such programmes; so all Pn-programmes that do not approach or meet the c-axis, in (c,x) space, attain zero consumption in finite time so violating Eq.(28) i.e. there is no T for which they can be optimal. If $\underline{n} \leq 0$ then zero consumption is attained only asymtotically. This implies that utility eventually becomes and remains negative. So if $\underline{n} < 0$, and therefore x becomes arbitrarily large, such programmes are dominated by switching, at some suitable time when x is sufficiently large, to a stationary programme for which $q = 0$ and so $c > c_o$. If $\underline{n} = 0$ then again such programmes can be dominated: by switching to a programme which approaches the origin.

In fact all the programmes considered above are inefficient for the resource is not exhausted. Therefore if the horizon is infinite the optimal programme, if one exists, must approach the origin; both c and x go to zero. Consider the subset of Pn-programmes with this property. Observe that:

$$S \equiv dx/dc = - u''(c)\,[c + x\,\underline{n}]/\rho u'(c) > 0$$

$$\partial S/\partial x = -\underline{n}\; u''(c)/\rho u'(c) \gtreqless 0 \leftrightarrow \underline{n} \gtreqless 0$$

so there will be a unique such programme if and only if $\underline{n} \leqslant 0$. If $\underline{n} > 0$ it is easily verified that any programme which approaches the origin can be dominated by switching to another that approaches the origin; so no optimal solution exists when $T = \infty$.

Now;

$$\frac{d[e^{-\Delta}px]/d\Delta}{e^{-\Delta}px} = \frac{\underline{n}}{\rho - \underline{n}} - \frac{dx}{d\Delta}$$

$$\rightarrow \underline{n}\ (\rho - \underline{n}), \text{ from above, as } \Delta \rightarrow \infty \ .$$

Therefore, if $\underline{n} < 0$, the sufficiency conditions are satisfied and there is a unique optimal programme when $T = \infty$. If $\underline{n} = 0$ the sufficiency conditions may not be satisfied; but then, when a sufficiently small $x(\Delta)$ has been attained, the model is equivalent to Gale's and so we know that existence depends on the behaviour of u (.) as $c \rightarrow 0$. This is summarised by:

Proposition 1

If $T = \infty$ then $\underline{n} \leq 0$ is necessary, and $\underline{n} < 0$ is sufficient, for the existence of an optimal programme. If $\underline{n} = 0$ existence depends on further information on the behaviour of u(.).

Proposition 2

If $T = \infty$ the optimal programme, if one exists, is the unique Pn-programme for which:

$$\lim_{\Delta \rightarrow \infty} c(\Delta) = \lim_{\Delta \rightarrow \infty} x(\Delta) = 0$$

In the context of Gale's model in which $\underline{n} = \bar{n} = 0$ and $\rho > 0$ it may seem counter-intuitive that existence depends on such precise knowledge of u(.). Proposition 1 indicates, however, that this is simply the borderline case between non-existence of an optimal programme when $\underline{n} > 0$ and existence when $\underline{n} < 0$.

If x_0 is large the optimal programme, at least for an initial phase, calls for allowing the largest number ($n = \pi \geq 0$) to enjoy high positive utilities (in fact $c > c^* > c_o$) but at the expense of an ever declining level of consumption and resource-availability per-capita. But the resource must not be exhausted in finite time, all generations must have a share albeit a small one, so at some point in time population policy is reversed and population sizes minimised ($n = \underline{n} \leq 0$). Both c and x continue to decrease monotonically and the economy asymtotically decays: later generations will be below the welfare subsistence level. Observe that the switch in population policy is unique for $dq/d\Delta = - xdp/d\Delta$, and occurs when $q = 0$ which is well before the time at which utility becomes negative.

Now consider Pn-programmes for which, at some finite time T, $x(\Delta(T)) = 0$. As $dc/dx > 0$ along such programmes it can be denoted $c^T = c^T(x)$ say. For given x, c^T is a continuous and strictly increasing function of $\Delta(T)$, and so T, defined on $[0, \infty]$ i.e. there exists a unique Pn-programme such that $x(\Delta(T)) = 0$ at some finite time T. It is efficient, and so optimal, relative to the planning horizon T.

Proposition 3

It T is finite there exists a unique T-optimal programme $c^T = c^T(x)$ say with $dc^T/dx > 0$ and, for any given x, $dc^T/dT < 0$. An obvious corollary, which is in accord with the results of the preceding sections, is:

Proposition 4

Amongst those Pn-programmes which are feasible for at least T periods the T-optimal programme is uniquely identified by maximising initial per-capita consumption (T finite or infinite).

There will exist a maximum T* such that if $T \leq T^*$ the T-optimal programme requires $n = \bar{n}$ and $c \geq c^*$ for all $t \leq T$. If $T > T^*$ the T-optimal programme has the same qualitative properties as the infinite horizon optimal programme except that per-capita consumption is everywhere higher. Once again there can be at most one change in the direction of population policy. Furthermore the qualitative results in no way depend on the welfare subsistence level.

(iii)"Doomsday"

The second step of the optimisation procedure is to determine the date at which all life should be ended i.e. the planning horizon T is to be determined as part of the analysis.

Recall that $c = c^T(x)$ is the unique Pn-programme for which 'x' is zero at the terminal date and so is optimal for this particular horizon; the infinite horizon optimal programme being the limit of these functions as $T \to \infty$. Choice of an optimal horizon, or survival time, is equivalent to choosing a particular function $c^T(x)$.

The argument will proceed in terms of time 't' and all variables will be implicitly functions of t. The symbol Δ can then be reserved for total differentials. Recall that W_T is the implied value of the social welfare function;

$$W_T = \int_0^T e^{-\rho t} L^{\alpha} u(C/L)\, dt$$

Now consider a feasible peturbation, of any T-optimal programme, which is efficient in the sense that the resource is exactly depleted at the terminal date of this programme. The horizon is to be extended in response to peturbations $(\Delta C, \Delta L)$ on the interval $[0,T]$. Therefore:

$$\Delta W_T = e^{-\rho T} L_T^{\alpha} u(c_T)\,\Delta T + \int_0^T e^{-\rho t} L^{\alpha-1} \{[\alpha u(c) - cu'(c)]\,\Delta L + u'(c)\,\Delta C\}dt$$

$$= e^{-\rho T} L_T^{\alpha} u(c_T)\,\Delta T$$

$$+ e^{-\rho T} L_T^{\alpha-1} u'(c_T) \int_0^T \{ [\alpha u(c) - cu'(c)]/u'(c).\ \ \Delta L + \Delta C\}dt$$

for $e^{-\rho t} L^{\alpha-1} u'(c)$ is constant along all Pn-programmes and so along all T-optimal programmes (see Eq. (18) or formulate the model in extensive form). Also:

$$X(0) = \int_0^T Cdt$$

so;

$$0 = C_T\,\Delta T + \int_0^T \Delta Cdt$$

which implies:

$$\int_0^T \Delta C.\ dt = -\ C_T\,\Delta T = -L_T\,c_T\,\Delta T.$$

After substituting this restriction, one obtains:

$$\text{(29)}\qquad \begin{aligned}\Delta W_T &= e^{-\rho T} L_T^{\alpha} [u(c_T) - c_T u'(c_T)]\,\Delta T \\ &+ e^{-\rho T} L_T^{\alpha-1} u'(c_T) \int_0^T [\alpha u(c) - cu'(c)]/u'(c).\ \Delta L\ dt\end{aligned}$$

Before attempting to apply this result it is useful to recall certain information. Eq. (22) implies:

$$c_T = c^T(x_T) = c^T(0) = c^*$$

if $T(\alpha)$ is finite. But W_T depends on x_o and, for given x_0, there may not exist a T-optimal programme with this property (see Figures 19 and 20 for which $\rho + \underline{n}\,(1-\alpha) \leq 0)$ and then $\hat{T}(\alpha) = \infty$. Even if there is such a programme it may not be unique (see Figure 20 and observe that when x_o is sufficiently small the programme in question may "double back" on itself). Along $q = 0$:

$$dq/dt \gtrless 0 \leftrightarrow \alpha u(c) - cu'(c) \lesseqgtr 0$$

and $(\tilde{c},\tilde{x})$ is the point at which $dq/dt = q = 0$.

Finally:

$$\partial L^{\alpha} u(C/L) \,/\, \partial L = 0 \implies \alpha u(c) - cu'(c) = 0$$

$$\implies c = \tilde{c}.$$

i.e. $\tilde{c}$ and c^* are related, in a specific manner, and $\alpha = 1$ implies $\tilde{c} = c^*$.

Let $c = c^{T^*}(x)$ be the unique Pn-programme, which meets $c = c^*$ and $x = 0$ at time T^*; T^* may not be unique. Restrict attention to T-optimal programmes.

Assume:

$$\underline{\rho + \underline{n}(1-\alpha) > 0}$$

Then $c^T(x)$ is a decreasing function of T and is defined for all x.

If $T < T^*$ then $c_T > c^*$ and $u(c_T) - c_T\, u'(c_T) > 0$

Increase the horizon to obtain $c = c^{T'}(x)$, with $T' \leq T^*$, so $c^T(x) > c^{T'}(x) \geq c^{T^*}(x)$. If this programme has the property $n_t = \bar{n}$ for all t, as is certainly the case if $\alpha = 1$ and sometimes the case if $\alpha < 1$, then $\Delta L = 0$ on the interval $[0, T]$. Therefore $\Delta W_T > 0$. If it does not have this property, so $\alpha < 1$, then move along the programme $c = c^T(x)$ until the terminal phase, in which $\alpha u(c_t) - c_t\, u'(c_t) < 0$ for all $t \geq s$ say, is attained and then switch to the programme $c = c^{T'}(x)$; $\Delta T > 0$ and $T' \leq T^*$. Then $\Delta L = 0$ on some interval $[0, \tau]$ and $\Delta L < 0$ on

$(\tau,T]$ so again $\Delta W_T > 0$. Now assume $T > T^*$ so $c_T < c^*$ and $u(c_T) - c_T u'(c_T) < 0$. At some time τ say it will be feasible to switch to a programme $c = c^{T'}(x)$, $\Delta T < 0$ and $T' \leq T^*$, which has the property $n_t = \underline{n}$ for all $t \geq \tau$. Then again $\Delta L = 0$ on $[0,T]$ and $\Delta W_T > 0$.

This indicates that a programme which does not have the property $c_T = c^*$ can always be dominated by a suitable switch in which the horizon is increased if $T < T^*$ and vice versa. Our search is among T- optimal programme, $T \in [0,\infty]$, and continuity of W_T in T for given x_0 indicates that an optimal $T = \hat{T}(\alpha)$ exists. So the unique programme $c = c^{T^*}(x)$ is optimal, i.e. $\hat{T}(\alpha) = T^*$.

<u>Proposition</u>

Suppose $\rho + \underline{n}\ (1-\alpha) > 0$ and n_t, c_t <u>and</u> T are the subject of choice, Then the optimal programme is the unique Pn-programme, defined for all x, for which $c_{\hat{T}} = c^*$. The resource is exhausted at time $\hat{T}$ and $dc/dt < 0$ for all t. If $\alpha = 1$ then $n_t = \bar{n}$ for all t. If $\alpha < 1$ then $n_t = \underline{n}$ at least in the terminal phase.

Observe that for any x_0, due to the possibility of negative population growth, it is feasible for society to survive for all time; but nevertheless optimality requires $\hat{T}$ to be finite. Also, whilst classical utilitarianism (α=1) requires population growth to be at its upper bound, the result is not robust; for any degree of "congestion externality" (α<1) will necessitate negative population growth in a terminal phase. In addition, if $\alpha < 1$, there is at most one switch in the direction of population policy; from $\bar{n}$ to $\underline{n}$. Finally, irrespective of 'α', all generations enjoy positive utility.

Suppose $\alpha = 1$, $\bar{n} > 0$, and the utility function has constant elasticity $\sigma < 0$. Then:

$$\dot{C}/C \equiv \dot{c}/c + n_t$$

$$= \dot{c}/c + \bar{n} = \bar{n} + \rho/\sigma$$

i.e. along the optimal programme $n_t = \bar{n}$ and if $\bar{n}$ is high, so $\rho + \sigma\bar{n} < 0$. then resource extraction increases with time and vice-versa. If now 'α' decreases from unity then:

$$\dot{C}/C = n_t + [\rho + n_t(1-\alpha)]/\sigma$$

so, if $\rho + \underline{n}(1-\alpha) > 0$, resource extraction decreases with time at least in the terminal phase. Initially, however, $n_t = \bar{n}$ and resource extraction may still increase; but at a lower rate than when $\alpha = 1$.

Now consider the behaviour of c_t over time and length of the optimal horizon. If $\rho = 0$ and $\bar{n} = \underline{n} = 0$ it is known [20] that $c_t = c^*$ for all time and $\hat{T} = x_0/c^*$. The introduction of positive discounting indicates that $\dot{c}/c = \rho/\sigma$ so initial consumption increases, but terminal consumption remains at c^*, so "doomsday" must be advanced i.e. $\hat{T} < x_0/c^*$. Now suppose it is possible to set $n_t = \bar{n} > 0$; the possibility of a population policy has been introduced. Then if $\alpha = 1$, so $n_t = \bar{n} > 0$ for all time, again $\dot{c}/c = \rho/\sigma < 0$. However, for given (c,x), dc/dx is now smaller. As c_T is anchored at c^* this indicates that initial per-capita consumption is reduced and, as the rate of decrease of per-capita consumption is the same as before, this indicates that c_t has been reduced for all time and again "doomsday" has been advanced. The next step is to let α decrease from unity; then per-capita consumption will decrease from a higher

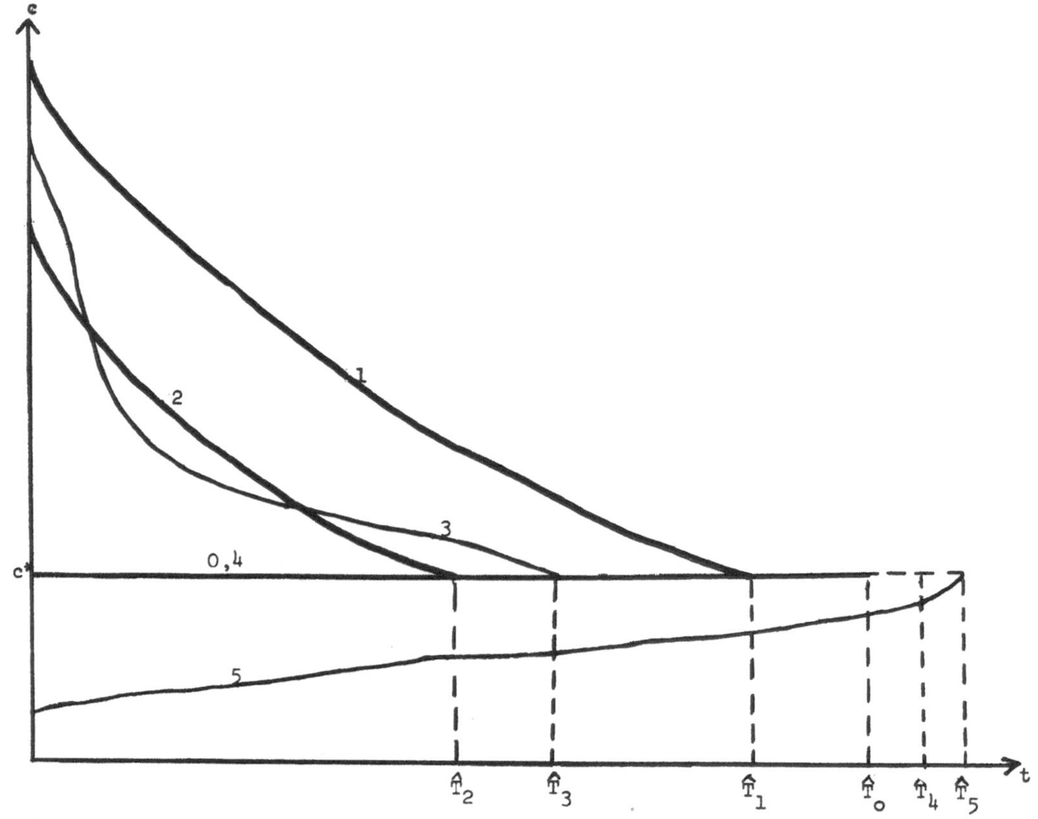

$\hat{T}_5 > \hat{T}_4 > \hat{T}_o > \hat{T}_1 > \hat{T}_2$; position of $\hat{T}_3$ indeterminate

$\forall x_o$:
- $\hat{T}_o$: $\rho = n = o$.
- $\hat{T}_1$: $\rho > 0$, $n = 0$.
- $\hat{T}_2$: $\rho > 0$, $\bar{n} > 0$, $\alpha = 1$.
- $\hat{T}_3$: $\rho > 0$, $n_o = \bar{n} = n_T$, $\alpha < 1$ but $\rho + \underline{n}(1-\alpha) > 0$.

$x_o < x^m$:
- $\hat{T}_4$: $\rho + \underline{n}(1-\alpha) = 0$, $\rho > 0$, $x^m = -c^*/\underline{n}$
- $\hat{T}_5$: $\rho + \underline{n}(1-\alpha) < 0$, $\rho > 0$, $x^m < -c^*/\underline{n}$

Figure 21: the optimal horizon

level; at a faster rate in the initial phase (when $n_t = \bar{n}$), and at a slower rate in the terminal phase (when $n_t = \underline{n}$). The existence of the "congestion" externality has created a bias in favour of earlier generations at the expense of later generations. Unfortunately it does not appear possible to determine if $\hat{T}$ has increased or decreased because the time at which n_t switches is not known. It is conjectured that the results would depend on x_o. These results are summarised in Figure 21.

An analogous argument to that used earlier shows that when α has decreased sufficiently, so that

$$\underline{\rho + \underline{n}\,(1-\alpha) = 0}$$

the optimal policy necessitates a finite $\hat{T}$ only if $x_0 < \underset{\sim}{x}$ where $\underset{\sim}{x} = -\overset{*}{c}/\underline{n}$ (see Figure 19); otherwise $\hat{T}$ is infinite. The optimal policy requires:

$$\left.\begin{aligned} & x_o < -c^*/\underline{n} \Longrightarrow c_t = c^* \\ \tilde{\tilde{x}} \geq \quad & x_o \geq -c^*/\underline{n} \Longrightarrow c_t = -x_o\underline{n} \end{aligned}\right\} \Longrightarrow n_t = \underline{n} = -\rho/(1-\alpha)$$

In particular so long as x_o $\tilde{\tilde{x}}$, per capita consumption is always less than $\tilde{c}$ ($\tilde{c} \ni \partial L^{\alpha}\, u\,(C/L)/\partial L = 0$) and the population growth rate is always at its lower bound. In sharp contrast, if $x_o > \tilde{\tilde{x}}$, then per capita consumption monotonically decreases but is bounded below by $\tilde{c}$, and the population growth rate is set at its upper bound. The point $(\tilde{c}, \tilde{x})$ is attained in finite time and is maintained, i.e. $\hat{T} = \infty$, so $n_t = \underline{n} = -\rho/(1-\alpha)$.

The most important points to emerge are summarised by:

<u>Proposition</u>

Assume $\rho + \underline{n}\,(1-\alpha) = 0$. Then if $x_o \geq \underset{\sim}{x}$ the optimal horizon $\hat{T}_4$ is infinite; but if $x_o < \underset{\sim}{x}$ it is finite in which case

$c_{\hat{T}} = c^*$.and, trivially, $\hat{T}_4 > \hat{T}_1$. Also:

$$n_t = \begin{cases} \bar{n} \\ \underline{n} \end{cases} \leftrightarrow \quad c \gtrless \tilde{c} \quad \leftrightarrow \quad x \gtrless \tilde{x}.$$

In particular, the optimal programme depends on initial per-capita stocks.

Finally, assume:

$\rho + \underline{n}\,(1-\alpha) < 0$, $(\tilde{c},\tilde{x})$ exists .

Then $c = c^{T^*}(x)$ is defined only for $x \leq \underset{\sim}{x}$ say where $\underset{\sim}{x} < -c^*/\underline{n}$. Suppose $x_o < \underset{\sim}{x}$. Then there are two programmes $c = c^{T_1^*}(x)$ and $c = c^{T_2^*}(x)$, $T_1^* < T_2^*$, for which $c_{T_2^*} = c_{T_1^*} = c^*$. They are identical on the interval $[0, x_o)$. It is easily verified that T_1^* is a local maximum, and T_2^* a local minimum, of W_T with respect to T. If T is increased beyond T_2^* then W_T continually increases; for $u(c_T) - c_T\, u'(c_T) > 0$, $\alpha u(c_t) - c_t\, u'(c_t) < 0$ for all t, and $\Delta\, L < 0$ on a terminal interval and zero otherwise. Therefore the critical issue is the sign of $W_\infty - W_{T_1^*}$ and this is indeterminate. In general there will exist an $x_c \in [0, \underset{\sim}{x}]$ which plays the same role as $\underset{\sim}{x}$ in the above proposition; the result is therefore quite robust.

<u>Proposition</u>

Assume $\rho + \underline{n}\,(1-\alpha) < 0$. Then the preceding proposition remains valid if $\underset{\sim}{x}$ is replaced by x_c. Furthermore $x_o \leq x_c \Rightarrow \hat{T} = \hat{T}_5$ say and $\hat{T}_5 > \hat{T}_4$. (see Figure 21).

It should be observed that as α decreases both $\tilde{c}$ and $\tilde{x}$ continually increase and $x_c \to 0$. For 'small' α and 'almost all' $x_o < \tilde{x}$ the optimal policy requires (c_t, x_t) to increase monotonically to $(\tilde{c},\tilde{x})$, $n_t = \underline{n}$ and $\hat{T} = \infty$. In fact, for all α below some critical level, $(\tilde{c},\tilde{x})$ does not exist (for existence requires $q = 0$ and $c = \rho x/(1-\alpha)$ and, as α decreases, $q = 0$ becomes steeper and $c = \rho x/(1-\alpha)$ flatter). In essence $(c,x) = (\infty,\infty)$ and the qualifications in the above results can be dropped.

This "average utilitarian" result is a complete reversal of the implications of classical utilitarianism; for then $n_t = \bar{n}$, $\hat{T}$ is finite and (c_t, x_t) monotonically decreases to $(\tilde{c},\tilde{x}) = (c^*,0)$.

Irrespective of the size of α, and so the sign of $\rho + \underline{n}(1-\alpha)$, if $x_o > \tilde{x}$ then the optimal policy requires that (c_t, x_t) monotonically decrease to $(\tilde{c},\tilde{x})$. If also $\rho + \underline{n}(1-\alpha) \leq 0$, then $n_t = \bar{n}$ until $(\tilde{c},\tilde{x})$ is attained at which point it is maintained and $n_t < 0$; the implications of classical utilitarianism remain valid (in essence for $\hat{T}$ is now infinite and $x(\infty) > 0$). Of course the price paid for this result is that $\tilde{x}$ itself increases as α decreases.

(iv) A "wealth" effect

In this section it is assumed that individualistic utility functions incorporate a wealth effect or conservation motive so:

$$u = u(c) \leftrightarrow u(c, x).$$

Such a formulation has been examined by Vousden [43] but under the assumption of a constant population; the interdependence of a wealth effect with a population policy, in the context of optimal exhaustible resource depletion, is now examined.

Some additional structure is imposed on the utility function; it serves to "tidy up" the analysis but is not strictly necessary for the results obtained. Specifically:

$$u_{cx} = u_{xc} = 0 \text{ for all } (c,x)$$

$$u_x(c,0) = \infty \quad u_x(c,\infty) = 0$$

Also assume $\rho > \alpha\bar{n} > 0 > \underline{n}$.

Recall that $q(\alpha) \equiv \alpha H - px$, $\theta^{\alpha}(c,x) \equiv \alpha u_x - \rho u_c$

and:

$$(30) \qquad \phi^{\alpha}(c,x) \equiv dc/d\Delta = \frac{u_c \, [\rho + n\,(1-\alpha)] - u_x}{u_{cc} \, [\rho - \alpha n]}$$

follows from Eq.(18), $p = u_c\,(c,x)$ and $u_{cx} = 0$. Then Eqs. (23) to (25) inclusive give the required information on these functions. It is summarised by Figure 22 which is now explained. The locus $q\,(\alpha) = 0$, for any $\alpha \in [0,1]$, has negative slope for points (c,x) that locate to the left of the $\theta^{\alpha}(c,x) = 0$ locus and vice-versa; it is specifically the introduction of the wealth effect that has lead to the former possibility. The $\theta^{\alpha}(c,x) = 0$ locus has positive slope for any α. Above the switching locus $q\,(\alpha) = 0$, q takes on positive values, and vice-versa.

The locus $\phi^{\alpha}(c,x) = 0$ coincides with $\theta^{\alpha}(c,x) = 0$ if $\alpha = 1$. As 'α' decreases the $q\,(\alpha) = 0$ locus shifts upwards but pivoted at $c^*(0)$. The $\phi^{\alpha}(c,x) = 0$ locus will still have positive slope, wherever it is defined, but now there is a segment of the switching locus that separates the $\phi^{\alpha}(c,x) > 0$ and $\phi^{\alpha}(c,x) < 0$ regions. Irrespective of 'α', $\phi^{\alpha}(c,x)$ is negative to the right of the $\phi^{\alpha}(c,x) = 0$ locus, and vice-versa.

Intuitively, to the right of the locus $\phi^{\alpha}(c,x) = 0$, $u_c\{\rho + n(1-\alpha)\} > u_x$, so the conservation motive is weaker than the preference for current consumption over future consumption, and therefore per-capita consumption falls over time. Finally notice that if $\rho + \underline{n}\,(1-\alpha) \le 0$ then $dc/d\Delta > 0$ everywhere in the $q\,(\alpha) < 0$ region; with

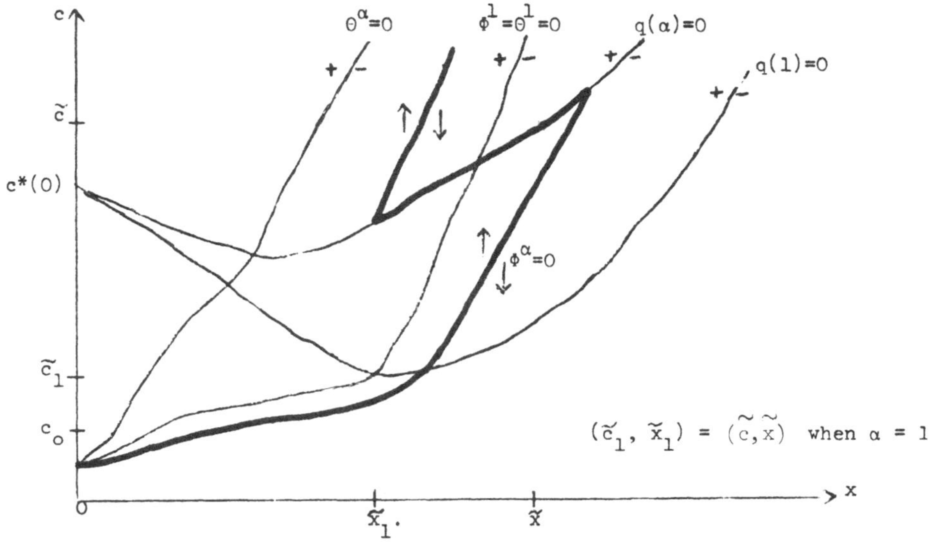

Figure 22: construction for conservation motive

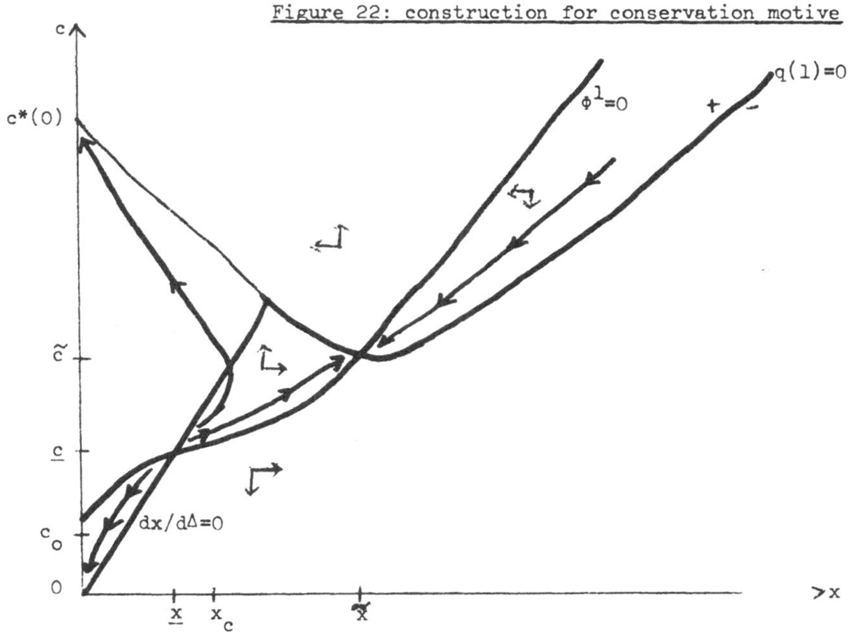

Figure 23: Conservation motive and α=1

the particular implication that there cannot then exist any Pn-stationary points in this region.

The welfare subsistence level c_o is now a function of x; so is c^* and $c^*(x) > c_o(x)$ for all x. Utility is always positive along the switching locus $q(\alpha) = 0$ for then $c \geq c^*(x)$ with equality only if $x = 0$.

The definition of $(\tilde{c},\tilde{x})$ in the previous section must now be modified and in such a manner that it plays the same role. This is so if $(\tilde{c},\tilde{x})$ satisfies:

$$\partial L^{\alpha} u(C/L,\ X/L)/\partial L \equiv L^{\alpha-1} [\alpha u - cu_c - xu_x] = 0$$

and;

$$q(\alpha) = 0.$$

In effect, $(\alpha u - cu_c)$ transforms to $(\alpha u - cu_c - xu_x)$, and with this modification it is the case that the expression derived above, which indicates the change in welfare achieved by extending the horizon of a T-optimal programme, remains valid. Perhaps more important is that Eq.23 (iii) indicates $(\tilde{c},\tilde{x})$ to be the only point on the $q(\alpha) = 0$ locus for which $dq(\alpha)/d\Delta = 0$. If $\alpha = 1$ then $(\tilde{c},\tilde{x})$ corresponds to the intersection of the two loci $q(\alpha) = \Phi^{\alpha}(c,x) = 0$ and if $\alpha < 1$ then $(\tilde{c},\tilde{x})$ lies to the right of the $\theta^1(c,x) = \Phi^1(c,x) =$ locus. Therefore both $\tilde{c}$ and $\tilde{x}$ increase when α decreases. Also $(\tilde{c},\tilde{x})$ is a Pn-stationary point if and only if the locus $dx/d\Delta = 0$ does not meet the $q(\alpha) = 0$ locus at some $x > \hat{x}$. For sufficiently low α, and $\rho + \underline{n}(1-\alpha) < 0$, the point $(\tilde{c},\tilde{x})$ may not exist.

An alternative interpretation of $(\tilde{c},\tilde{x})$ is of interest. Suppose the resource depletion profile, and so also the terminal date, is given and that at any moment of time there is common ownership of the remaining resource stock between members of the current generation. If population size L is to be chosen optimally, so "effective" population size is L^{α}, then the social marginal costs and benefits consequent on introducing an additional member into society must be equal. The additional member will consume an amount 'c' of the resource so attaining utility u(c) and society benefits to the extent:

$$u(c)\,\Delta L^{\alpha} = \alpha L^{\alpha-1}\,u(c)\,\Delta L$$
$$= \alpha L^{\alpha-1}\,u(c)\,.$$

The marginal cost to society consists of two components; the direct effect which is the amount 'c' of the resource consumed by this member and the indirect effect which is this member's share of the resource, $x = X/L$, evaluated at the price of X in terms of the numeraire C i.e. u_X/u_C. The total effect is $c + x\,u_X/u_C$ and this must be evaluated at the correct scarcity price, i.e. the "effective" population size L^{α} times the marginal utility of a unit of consumption u_C. So at the margin:

$$\alpha L^{\alpha-1}\,u(c) = L^{\alpha}\,u_C\,[c + x\,u_X/u_C]$$

or;

$$(31) \qquad u(c) = [cu_c + xu_x]$$

This is the counterpart of Meade's rule in the context of exhaustible resources.

Now suppose the resource depletion profile can also be varied. Then both Eq. (30), if $x > 0$, and Eq. (31) are necessary for optimality. However, as X_o is given but L controllable for all $t \geq 0$, x_o is controllable and $x = \tilde{x}$ can therefore be attained instantaneously. But at $(\tilde{c},\tilde{x})$ not only is Eq. (31) satisfied but also $dc/d\Delta = 0$. Because 'n' is not constrained $(\tilde{c},\tilde{x})$ can be maintained into the indefinite future. By the transversality conditions it is optimal to do so.

In summary, if population sizes rather than population growth rates, are the subject of choice, then the optimal policy is to set $(c,x) = (\tilde{c},\tilde{x})$ for all time (see Figures 18 to 23 inclusive excepting 21). There are two qualifications to be made. If $\alpha = 1$ and $u_x \equiv 0$ then $(\tilde{c},\tilde{x}) = (c^*,0)$ and the optimal policy calls for consuming everything today and terminating immediately. It is also possible that, for sufficiently low α, $(\tilde{c},\tilde{x})$ does not exist; in which case it is possible that no optimal policy exists.

The above results give normative significance to the programme $(\tilde{c},\tilde{x})$ which, irrespective of whether there is a "wealth effect" or not, plays an important role when the population growth rate is controllable between given upper and lower bounds.

Returning to the full model consider the possible existence of Pn-stationary points, $(\underline{c},\underline{x})$ say, with $\underline{x} \in (0, \tilde{x})$ and locating in the $q(\alpha) < 0$ region. If $\rho + \underline{n}(1-\alpha) \leq 0$ there are none but otherwise there may be many. In the latter case, to simplify the exposition, it is assumed that there is at most one such point; it is a simple matter to modify the results if this is not so. As $(\underline{c}, \underline{x})$ must lie on the ray $\underline{c} = -\underline{n}\,\underline{x}$, and also $u_x/u_c = \rho + n(1-\alpha)$, an assumption sufficient to guarantee this is that the utility function exhibit a linear "income expansion" path in the relevant region.

Now suppose $\alpha = 1$. The relevant information in Figure 22 is duplicated in Figure 23. Only the Pn-programmes for which either $T = \infty$

or $c_T = c^*(0)$ have been illustrated for these conditions are necessary for optimality if the horizon is chosen optimally. The programmes are labelled $c = c^{\infty}(x)$ and $c = c^{T^*}(x)$ respectively. It is implicit in Figure 23 that:

(i) $\underline{n}$ is sufficiently negative that $(\tilde{c},\tilde{x})$ is a stationary point and

(ii) $u_{cc}(0,0)\,[\rho + \underline{n}\,(1-\alpha)] = u_c(0,0)\,\rho > u_x(0,0)$

i.e. the discount motive is relatively strong in the sense that it exceeds the conservation motive in a neighborhood of the origin. The implication is that $dc^{\infty}(x(\Delta)/d\Delta < 0$ for small x. From (i) and (ii) the existence of $(\underline{c},\underline{x})$ is ensured.

Suppose $x_o < \underline{x}$ then along the programme $c = c^{\infty}(x)$ utility eventually becomes and remains negative; so, as the horizon is the subject of choice, this can be dominated by switching to $c = c^{T^*}(x)$. If x_o is close to $\tilde{x}$ clearly $c = c^{\infty}(x)$ is optimal; for $c = c^{T^*}(x)$ is not defined. As W_T is strictly increasing in x_o there is a critical level of x, x_c say, such that:

$$\left.\begin{array}{l} x_o < x_c \implies c = c^{T^*}(x) \\ x_o > x_c \implies c = c^{\infty}(x) \end{array}\right\} \quad \text{is optimal}$$

Again T* is not unique. There are two solutions, T_1^* and T_2^* say, and the optimal $\hat{T}$ is:

$$\min\,(T_1^*,\ T_2^*)$$

for only then is $dx/d\Delta$ one-signed.

The optimal programme is illustrated in Figure 24. The conservation motive has introduced the possibility of "doomsday" being postponed into

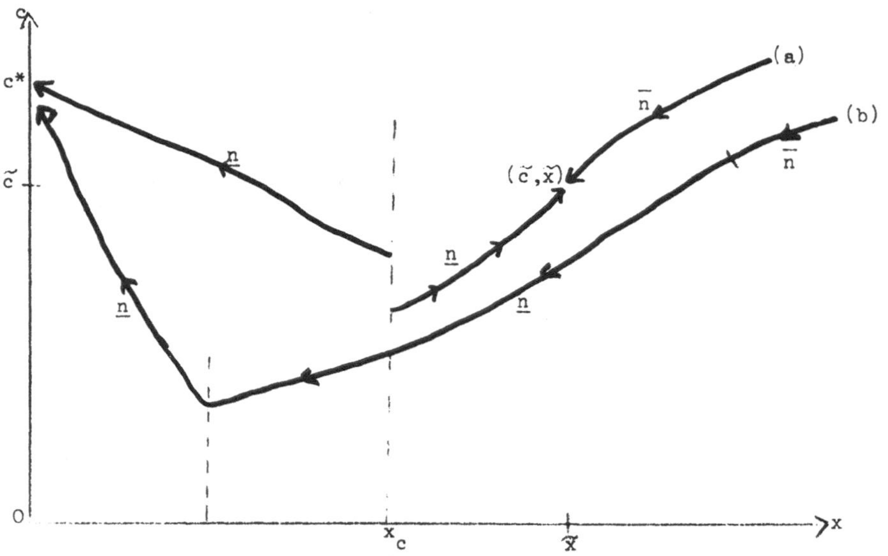

(a) $\underline{n} < 0$ and very small; $(\tilde{c},\tilde{x})$ a stationary programme

(b) $\underline{n} < 0$ but close to zero; $(\tilde{c},\tilde{x})$ not a stationary programme.

Figure 24: Conservation motive and optimal programmes

the indefinite future; at least if $x_o > x_c$. There is a threshold theory of growth; the critical levels being x_c and $\tilde{x}$. The programme $(\tilde{c},\tilde{x})$ is attained in finite time, if $x_o > x_c$, and from then on maintained. Population growth will be at its maximal rate during the transition to $(\tilde{c},\tilde{x})$ if $x_o > \tilde{x}$ and vice-versa. If $x_o < x_c$ then $n = \underline{n}$, per-capita consumption always increases, but now at the expense of potential generations after time T_1^*.

The optimal programme has the same qualitative features as in the absence of a conservation motive providing α is low. (see Figure 20). Furthermore, variations in α , given the existence of a conservation motive, and assumption (i), imply optimal programmes with the same qualitative features relative to $(\tilde{c},\tilde{x})$ as when $\alpha = 1$. But $(\tilde{c},\tilde{x})$ will increase as α decreases and, as $\underline{x}$ will decrease, this suggests x_c decreases also. In short the conservation motive has added a greater degree of robustness to the results. Even if assumption (ii) is removed the optimal programme has the same form except that x_c may now be zero.

If assumption (i) is relaxed but (ii) maintained then, irrespective of α , $\hat{T}$ is finite. If x_o is very large then initially $n = \bar{n}$. At some (c,x), for which $c > \tilde{c}$ and $x > \tilde{x}$, the population growth rate switches to its minimal rate $\underline{n}$ and is maintained at $\underline{n}$. Per-capita consumption first falls and then increases, due to the conservation motive, in the terminal phase; with this latter modification the optimal policy is qualitatively the same as in the absence of a conservation motive when 'α' is large. This result is invariant, when neither (i) or (ii) hold, if $u(\underline{c},\underline{x}) \leq 0$; otherwise further quantitative information is required to determine the optimal time horizon. The optimal programme is again illustrated in Figure 24.

As a final but important modification suppose $\underline{n} \geq 0$. Then there are no P_n-stationary programmes. An optimal programme will exist, however, and will have the form indicated by (b) in Figure 24; naturally the optimal horizon will be finite.

In essence this model serves to focus our attention on the case where the resource is 'essential' in an extreme sense; for there can be no consumption without it. The model should be extended to allow for possible substitution between this resource and a substitute resource which might be discovered at some later date. A further source of substitution is between the resource and the capital stock. While such issues have been considered, in depth, in the exhaustible resource literature, the population profile is always assumed exogenously determined so eliminating from the analysis the kinds of questions considered here. A more complete analysis is of importance.

6. An Alternative Framework.

The approach used in the previous sections, to determine a normative theory of population and savings, was to choose non-negative sequences $\{C_t\}$ and $\{L_t\}$, from some feasible set of such sequences, so as to maximize a social welfare function which in general takes the form:

$$W = G(C_o, C_1, \ldots C_t, \ldots; L_o, L_1, \ldots L_t, \ldots).$$

Now it was argued that our concern should be with the size and distribution of the utilities resulting from the pair of sequences $\{C_t\}$ and $\{L_t\}$ in which case the social welfare function takes the 'utilitarian' form:

$$W = W(u(c_o), u(c_1), \ldots, u(c_t), \ldots; L_o, L_1, \ldots, L_t, \ldots).$$

Such a formulation is too general to provide much guidance so only certain special cases of this function were considered; namely, W_A, W_B and W_C. Of course there are alternative, hopefully morally justifiable, restrictions of this W-function. However, the question has been asked [7] as to whether one can devise persuasive moral arguments that obligate the present generation to put into effect any program based on the maximisation of some particular W-function that attaches positive weights to the utilities of potential future generations.

The assumption that the utility of the representative individual depends only on his own per-capita consumption level will be retained. The size of generation zero is given and this generation must decide how much to consume, and so how much capital to bequeath to the next generation, and also the size of the next generation i.e. L_1. Now if future population sizes were not subject to choice but were given then there are certainly moral arguments for persuading the present generation to consider the utilities of future generations when deciding how much to consume today and so how much to save for the future; the claims of those who do not now exist, but definitely will exist, must be considered. But here the problem is not to determine an optimal savings program subject to a given population profile; rather it is to determine an optimal joint population/savings program i.e. population itself is controllable. In such circumstances one feels that there is a fundamental difference between the claims of the 'potential born' and those that are presently alive. For example one could ask why an individual would be morally obliged to applaud the creation of a child in the interest of increasing total utility if the birth of this child implied even a small loss to his own welfare. Taking the argument a step further one could ask if there are any moral reasons why the present generation should not consume the total output it produces and also produce no offspring ($L_1 = 0$).

A possible source of such moral arguments is utilitarianism and in a recent paper Narveson [30] has considered whether or not utilitarianism can provide an affirmative answer to any of these questions. He argues that according to broad utilitarian tenets all obligations, and all moral reasons for doing anything, are grounded upon the existence of persons who would benefit or be injured by the effects of one's actions. Utilitarianism requires us to find cause for concern in any situation where the well-being of some actual human being is depressed or destroyed or simply not considered. Refer to the well-being of a person when born as the direct effect of creating a new person. Also, refer to the alteration in well-being that his existence will cause on others as the indirect effect. For the purpose of exposition this indirect effect is assumed negligible; the inclusion of non-negligible indirect effects can only strengthen the argument. Now consider the birth of a happy child. One would neither have benefited nor injured anyone by making the decision not to give birth to the child; for the child would not have existed prior to its birth. Therefore, whilst utilitarianism does not condemn the decision to create the child, it certainly does not dictate the moral desirability of doing so. On the other hand utilitarianism does prohibit the creation of a child who one knows is going to be miserable; since after his birth one can refer to a person on whom one has inflicted misery. In short Narveson is arguing that utilitarianism gives rise to a duty not to have children, if it is known they will be miserable, but it can never give rise to a duty to have them even if it is known they will be happy. There is a fundamental asymmetry in the treatment of obligations and this seems to accord with our moral intuition.

This line of reasoning indicates that utilitarianism does not prescribe the use of any so-called 'utilitarian' social welfare function as given above. For when a utilitarian must compare alternative $\{u(c_t)\}$ and $\{L_t\}$ sequences he must remember that these do not refer to actual people but only to people who, partly as a result of his decision, may or may not exist; so there is no compulsion on him to bother about the results of the comparison or to set L_1 different from zero. Setting $L_1 = 0$ may appear distasteful but this seems to be an aesthetic consideration rather than a moral one.

In particular it follows that it is wrong to think that the classical utilitarian social welfare function W_c, and so also the Sidgwick-Meade population theory, is implied by utilitarian premises. Such a welfare function embodies a particular ethical viewpoint; namely that the happiness of the 'potential born' can and should be traded off against the happiness of those presently existing. Therefore a decision must be made to create an unhappy child whenever the indirect effects outweigh this negative direct effect. This is not compatible with the utilitarian premises. But the essential point is that, even if this viewpoint is considered reasonable, it is not utilitarianism that can obligate the present generation to attach positive weights to the utilities of potential future generations.

If utilitarianism cannot show such a committment there are a number of dubious arguments that do. For example one can argue that it is selfish to consider only ourselves and our actual descendents, if we choose to have them, for the unborn-and-never-will-be born are unfairly ignored. The image is that of potential immigrants to a land of plenty left, instead, in an impoverished state. The mistake lies in treating persons who do not exist and will not exist, as a special sort of person whom we are in danger of overlooking. But the question at issue in population policy is whether there shall be further people; if we decide against it then there is nobody whose welfare is overlooked for it is precisely the result of our policy that there should be nobody.

Whilst other such arguments have been considered (and correctly dismissed by Dasgupta [9]) we will restrict our attention to re-considering the Rawlsian "original position" concept, which was used to provide some sort of justification for the "probabalistic" social welfare function W_B, in light of the utilitarian argument above. This concept provides a conceptual apparatus that captures the consideration "suppose I were in his circumstances" so providing content to our notion of a "just" population policy. The difficulty in applying this concept to the problem of optimal population sizes is that there presumably must be a well-defined 'he' for this consideration to be relevant; for, as has already been argued, morally one may confine one's concern to those who do exist or actually will exist. Therefore this concept does not seem to make sense in the context of choosing a population policy.

Neither Harsanyi [14] nor Vickrey [42] were concerned with the problem of optimum population when advocating the utilitarian doctrine via an 'original position' argument; both assumed that the size of the population was given. Also, whilst Rawls [34] applies the 'original position' concept to the problem of determining a just inter-generational savings program, once again population is assumed exogenously given. Now of course one can apply this concept to ask whether the savings policy is just in its treatment of the different generations implied by a joint population and savings program. However, the subject of our inquiry is broader; it is precisely the determination of this joint population savings program.

Notice also that the "original position" argument applied to the W_B - social welfare function requires that the utility index of generation g, for example, be weighted by the probability that a member of this "original position" will also be a member of this particular generation. But not only is it not known which generation he will belong to; it is also not known if he will in fact ever exist. Now the weights in our W_B - social welfare function are conditional probabilities, conditional that is on his present or future existence, whereas the relevant probabilities (as implied by the 'original position' argument) are unconditional. Therefore our probabilistic, or expected utility, interpretation and justification of the W_B - function is not strictly correct.

An additional drawback of the social welfare function approach used earlier is that the only indirect effect of the creation of a person is his effect on the average level of others. The rather common sentiment, namely that children add joy to one's own life and that one is also concerned with their well-being, is therefore ignored. An explicit inclusion of these added dimensions has been made by Spengler [40] and more recently by Mirlees [28]; but the problem with these studies is that they are restricted to a timeless framework and it is not clear how to extend them into an inter-temporal context. If one merely sums all potential utilities then once again one is vulnerable to the kinds of objections that Narveson raised.

It seems then that any social welfare function approach provides an un-promising framework within which to consider different savings population programs. An alternative framework, designed to be sufficiently open to allow for the possibility of moral and aesthetic discourse within it, has been provided by Dasgupta [7]. The aim is not to derive a definite rule for optimal population sizes but rather to derive a class of programs consistent with a minimal set of both human aspirations with a wide empirical base and certain a priori moral considerations. The problem with the Sidgwick-Meade formulation is that it advocates a too special moral theory; and one that can be easily questioned.

Now consider what sorts of moral ingredients this theory should contain. Our earlier discussion indicates that the framework should not indicate a moral obligation to produce children simply because they would be happy; but it should indicate an obligation not to produce them if one knows they would be miserable. By this constraint no one,during the pursuit of a policy implied by this framework,will be miserable and so no one can complain for having been born.

Also, if a policy emanating from this framework is to command ethical attention, it must be in some sense universalisable. The policy will be assumed universalisable in the Rawlsian sense [7] that each individual, during the pursuit of this policy, would realise that he would have been willing to abide by it even if he had instead been born as someone else. A solution concept that captures this universalisability property is an inter-temporal version of the Nash non-cooperative equilibrium concept and this will be used below.

To capture the sentiment that children add joy to one's own life it is assumed that an individual has a direct concern not only for his own consumption level but also the size of his family and the consumption level of the others in his family. Supposing that all individuals at all times have identical preferences, and for simplicity that each generation lives for one time period only and generations do not overlap, one may write (using a discrete time formulation):

$$U^{(t)} = U(c_t;\ c_{t+1};\ n_t)$$

where $c_t \equiv C_t/L_t$, $n_t \equiv L_{t+1}/L_t$, $U(\cdot)$ is strictly concave and differentiable, and $U^{(t)}$ is the utility of a member of generation t. The last argument of $U(\cdot)$ indicates that the individual is concerned with his family size and not with the size of the next generation as such. Whilst one may have various views on the size of future populations they are not central to our purpose and therefore not included in the formulation. It is assumed that 'n_t' is a continuous variable.

Purely as a normalisation device it is supposed that the unborn state yields zero level of utility and also that the utility function can take on both positive and negative values. In the event that zero population is reached in finite time, and so that the 'last' generation does not save for the future, assume $U(c_t, c_{t+1}, 0) = U(c_t, 0, 0)$. Also suppose that:

$$\partial U/\partial c_t \equiv U_1^{(t)} > 0 \quad \text{and} \quad \partial U/\partial c_{t+1} \equiv U_2^{(t)} > 0$$

i.e. the marginal utility of one's own consumption and the average consumption of one's offspring is positive.

It seems natural to suppose that there exists an $\bar{n}$ such that:

$$n_t \gtrless \bar{n} \Leftrightarrow \partial U/\partial n_t \equiv U_3^{(t)} \gtrless 0$$

and to avoid the optimality of a zero consumption level assume $\underset{c_t \to 0}{\text{limit}}\, U_1^{(t)} = \infty$.

The initial population size and capital stock, L_0 and K_0 respectively, are given and so k_0 is given. To complete the specification of the model one must add to the assumptions on the preference structure a statement as to the form of the production function. It will be assumed to be linear homogeneous, strictly concave and differentiable everywhere so:

$$Q_t = F(K_t, L_t) \equiv L_t\, f(K_t/L_t) \equiv L_t\, f(k_t)$$

Therefore if C_t is total consumption by generation t the basic accumulation constraint is:

$$K_{t+1} - K_t = L_t\, f(K_t) - C_t$$

Denote by s_t the over-all savings ratio of generation t so,

$$c_t = (1 - s_t)f(k_t)$$

and this constraint reduces to:

$$n_t k_{t+1} = s_t\ f(k_t) + k_t$$

Then the utility level of generation t will be:

$$U^{(t)} = U[(1-s_t)f(k_t);\ (1-s_{t+1})f\{\frac{1}{n_t}(s_t f(k_t) + k_t)\};\ n_t]$$

Now generation zero will choose s_o and n_o to maximise $U^{(o)}$ subject to the moral constraint that $U^{(1)} \geq 0$ if $n_o > 0$; the preferences of the representative member at t = 1 must be respected. But this constraint depends on s_1, n_1 and s_2. Similarly generation one would maximize $U^{(1)}$, with respect to s_1 and n_1, subject to the constraint that $U^{(2)} \geq 0$ if $n_1 > 0$; and this depends on the choice of s_2, n_2 and s_3. In short, the present generation must contemplate the complete sequence $((s_o, n_o), (s_1, n_1), \ldots (s_t, n_t) \ldots)$ of savings ratios and family sizes for all future generations. Such programmes constitute a subset of all feasible $\{(s_t, n_t)\}$ sequences. Notice that no ordering is provided over this subset.

This subset may be restricted by incorporating the concept of universalisability discussed earlier. There is a moral requirement that the chosen policy should be one to which all affected parties could ideally agree. Therefore, for a program $\{(s_t^*, n_t^*)\}$ to be judged optimal, it must have the property that each individual who comes into existence during the pursuit of this policy finds his utility level maximised if he pursues this policy provided that all others who are ever born during the pursuit of this policy also choose according to this policy.

Stating mathematically the two moral constraints imposed on this framework one has:

> for a feasible population savings programme $\{(s_t^*, n_t^*)\}$ to be judged optimal it must satisfy:
>
> (i) for all $t \geq 1$ if $n_{t-1}^* > 0$ then $U^{(t)} \geq 0$ and
>
> (ii) for all $t \geq 0$ if $L_t^* > 0$ then the pair (s_t^*, n_t^*) maximises $U^{(t)}$ on the assumption that for all $t' \neq t$ the representative member of generation t' chooses the pair $(s_{t'}^*, n_{t'}^*)$.

Consider a feasible programme $\{(\bar{s}_t, \bar{n}_t)\}$ which does not satisfy (ii). Then there must exist an individual, at some time τ say, for whom the choice of $(\bar{s}_\tau, \bar{n}_\tau)$ does not maximise $U^{(\tau)}$ if he supposes that the representative member of every other generation $t (t \neq \tau)$ chooses $(\bar{s}_t, \bar{n}_t)$. Then such a policy $\{(\bar{s}_t, \bar{n}_t)\}$ could not be morally justified by a representative member of generation zero; for he could be asked, and would be obliged to answer no, whether he would be willing to pursue the policy if he had instead been born at time τ. It is in this sense that (ii) captures the concept of universalisability.

Notice that in this universalisability argument which is presented to the present generation the claims of no potential persons have been introduced; for the present generation does have the option of setting $s_o = 0$ and $n_o = 0$. But this generation will consider all its options. If it chooses to produce offspring (i.e. $n_o > 0$), so the program it selects must satisfy (i) and (ii), then it must be because this programme yields a higher utility level $U^{(o)}$ than if n_o was set equal to zero. But this is true for all generations so this framework calls for considering only the claims of those who actually will be born.

To simplify the argument suppose that condition (i) is automatically satisfied at the optimum i.e. the constraint $U^{(t)} \geq 0$ for all $t \geq 1$ is non-binding. This amounts to assuming k_o to be 'large' and that individuals are not too fastidious about what they want out of life. Also suppose that $n_t \neq 0$ for any t at the optimum; to some extent this is justifiable if L_o is 'large'.

Then the above discussion indicates that generation t chooses (s_t, n_t) to maximise $U^{(t)}$ so, assuming an interior solution, $\partial U^{(t)}/\partial s_t = 0$ and $\partial U^{(t)}/\partial n_t = 0$. Therefore:

(a) $$U_1^{(t)} = (1 - s_{t+1})f'(k_{t+1})U_2^{(t)}/n_t$$

(b) $$n_t^2 U_3^{(t)} = (1 - s_{t+1}) \{s_t f(k_t) + k_t\} f'(k_{t+1})U_2^{(t)}$$

Using the capital accumulation constraint condition (b) takes the more transparent form:

(c) $$U_3^{(t)} = U_2^{(t)}(1 - s_{t+1})[f(k_{t+1}) - \{f(k_{t+1}) - k_{t+1}f'(k_{t+1})\}]/n_t$$

Condition (a) indicates that a member of generation t chooses his savings ratio s_t so as to equate the marginal gain with the marginal loss; it is the counterpart of the Ramsey rule. For if a member of generation t reduces his consumption by one unit his loss in utility is $U_1^{(t)}$. But this savings yields $f'(k_{t+1})$ additional units of output for his descendants who will each consume a proportion $(1 - s_{t+1})/n_t$ which is valued at the correct scarcity price $U_2^{(t)}$. Again, condition (c) expresses the desire for equality in the marginal gains and losses consequent on introducing one more descendant; it is the counterpart of the Meade rule. For the introduction of this descendant will add utility $U_3^{(t)}$ to a member of generation t. But average output of generation t + 1 will be lowered by:

$$[f(k_{t+1}) - \{f(k_{t+1}) - k_{t+1}f'(k_{t+1})\}]/n_t$$

i.e. by the proportion $1/n_t$ of the difference between the average and marginal products of labor. Then a fraction $(1 - s_{t+1})$ of this fall in average output will come from consumption and this will be valued by the member of generation t at $U_2^{(t)}$.

As $s_{t+1} < 1$ condition (c) implies $U_3^{(t)} > 0$ so the optimal family size $\hat{n}_t$ is less than $\bar{n}$ i.e. no generation will be required to bear the burden of a family size at which its marginal utility of the family size is negative. Alternatively, by condition (c), the direct effect and the indirect effect (of introducing an additional child into society) are both positive (and equal). This is necessitated by utilitarian considerations but is in contrast to the Sidgwick-Meade formulation which can generate a situation in which this indirect-effect is negative.

The optimal programme in this framework is an infinite sequence $\{(c_t, k_t, s_t, n_t)\}$ which satisfies the constraints imposed by the technology and also conditions (a) and (c). In general there will be an infinite number of such solution sequences (or inter-temporal non-cooperative Nash equilibria) so conditions (a) and (c) are necessary, but not sufficient, for the determination of a unique morally commendable population savings program. This is as it should be, however, for this framework is designed to be sufficiently open to allow for moral discourse within it and so only provides limits on the kinds of arguments that generate optimal population savings programs. As a consequence the two conditions provide only a partial ordering over the set of feasible programs. But the point is precisely that the framework admits of some flexibility and additional considerations could be introduced to make the ordering complete.

Many of these Nash equilibria will be dynamically inefficient and so will not merit consideration. But it is difficult to generate the total set of efficient Nash equilibria in a general context and so we will consider a particular example in order to gain further insight.

Suppose that:

$$U^{(t)} = \log c_t + \beta \log c_{t+1} + (\gamma n_t - \delta n_t^2) + B \text{ if } n_t > 0$$

$$= \log c_t + B \text{ if } n_t = 0$$

and also $f(k) = k^{\alpha}$ where $0 < \alpha < 1$. B is a constant. Now $U_3^{(t)} = \gamma - 2\delta n_t$ and it seems reasonable to assume

$$\gamma > 1, \ \bar{n} = \gamma/2\delta \doteq 2 \text{ and } 0 < \beta \leq 1 .$$

Restrict attention to steady state Nash equilibria. Then the time subscript can be dropped and one has:

$$c \equiv (1 - s)k^{\alpha}$$

and the accumulation constraint:

$$(n - 1)k = sk^{\alpha}$$

Conditions (a) and (c) reduce to:

$$n = \alpha\beta(1 - s)k^{\alpha-1}$$

$$\alpha\beta(1-s)k^{\alpha} = nc(\gamma - 2\delta n)$$

The first and last equations yield a quadratic in n:

$$2\delta n^2 - \gamma n + \alpha\beta = 0 \ .$$

It can be argued that it is the larger of the two solutions to this equation that should be considered and it is denoted by $\hat{n}$ so:

$$\hat{n} = \frac{\bar{n}}{2} + \frac{1}{4}\left[4\bar{n}^2 - \frac{16\alpha\beta\bar{n}}{\gamma}\right]^{1/2} > 1$$

Assume that $\gamma/2\delta \equiv \bar{n} > 4\alpha\beta/\gamma$ so $\hat{n}$ is real and a steady state solution does therefore exist. The second and third equations above may be solved for the steady state savings ratio, capital labor ratio and per-capita consumption if desired. But only the variable $\hat{n}$ will be considered here.

As $\hat{n} > 1$ the steady state optimal program requires a growing population. Also it is intuitively clear that the higher is α (the share of capital) and the higher is β (a generation's concern for its descendants consumption level) then the lower is $\hat{n}$. The equation for $\hat{n}$ also shows that precise bounds can be placed on $\hat{n}$; namely $\bar{n}/2 < \hat{n} < \bar{n}$.

Returning to the general framework suppose now that each individual not only has views about the size of his own family but that he has views on the size of future populations as well; a consideration alluded to earlier. So assume:

$$U^{(t)} = U(c_t, c_{t+1}, \ldots, n_t, n_{t+1}, \ldots) .$$

Also assume that $U^{(t)}$ is separable to the extent that:

$$U^{(t)} = V(c_t, c_{t+1}, n_t) + \sum_{\tau=0}^{\infty} n_t \, n_{t+1} \ldots n_{t+\tau} V(c_{t+\tau+1}) ,$$

for $t \geq 0$. Then this can be interpreted as the Sidgwick-Meade social welfare function, although with more structure, if all individuals are assumed to possess a common utility function V. In fact the Sidgwick-Meade W_c function, used in sections 3 and 4, had the more restricted form:

$$U(t) = V(c_t) + \sum_{\tau=0}^{\infty} n_t \, n_{t+1} \ldots n_{t+\tau} \, V(c_{t+\tau+1}).$$

If generation t's preferences are of this form then it can be shown that one of the solution sequences $\{(s_t, n_t)\}$, if one exists, will in fact be the Sidgwick-Meade optimal population savings policy evaluated at $t = 0$. Thus the classical utilitarian solution is contained within a set of solutions emanating from this framework in which no longer are the claims of the 'potential born' treated on a par with the claims of those that are definitely committed to being alive.

Suppose that the sequence $\{(s_t^*, n_t^*)\}$ is optimal in the sense of satisfying conditions (i) and (ii), or equations (a) and (c), above. The question can be asked whether the savings sequence $\{s_t^*\}$ is 'just' as between the generations implied by the population profile $\{n_t^*\}$. It has in fact been argued [8] that the Rawlsian original position argument applied to the profile $\{n_t^*\}$ would generate the savings sequence $\{s_t^*\}$ i.e., this pure savings policy is 'just'. Notice that in this argument the sequence $\{n_t^*\}$ is given; generations cannot choose what number of descendants to have for it has already been argued that the original position argument cannot be applied to problems of population policy. But it is certainly legitimate to ask if the savings policy implied by a joint population savings programme is 'just' relative to that population profile.

In conclusion, it is noted that the basic argument of this section hinges on whether the fundamental moral entity is agreeable consciousness or persons. The former view is consistent with the social welfare function approach. However, if one evaluates actions by their implications for the rights of individuals, in which case the social welfare function approach is dubious simply because the domain of aggregation is not well-defined, an alternative procedure must be offered to guide one's choice amongst different savings/population programmes.

REFERENCES

1. Arrow, K.J. and M. Kurz. Public Investment, the Rate of Return and Optimal Fiscal Policy. Published for Resources of the Future, Inc., by the Johns Hopkins Press, 1970.

2. Becker, G.S. "An Economic Analysis of Fertility", in Universities-National Bureau Committee for Economic Research, Demographic and Economic Change in Developed Countries. Princeton: Princeton University Press, 1960.

3. Chichilnisky, G. "Some Results on Competitive Optimal Growth in a Many Sector Economy", Journal of Mathematical Analysis and Applications, forthcoming.

4. Cooter, R. "Contractual Theory of Optimal Population Growth and Capital Accumulation", mimeo, Harvard University, 1973.

5. Dalton, H. "The Theory of Population", Economica, 1928.

6. Dasgupta, P.S. "On the Concept of Optimum Population", Review of Economic Studies, Vol. 36(3) No. 107, 1969.

7. ____________, "On Optimum Population Size", from Economic Theory and Planning; Essays in Honour of A.K. Das Gupta, edited by Ashok Mitka, Oxford University Press, Calcutta 1974.

8. ____________, "Some Problems Arising from Professor Rawls' Conception of Distributive Justice", Theory and Decision.

9. ____________, Blackburn, S., "The Ethical Foundations of Population Policies".

10. ____________, Heal, G.M., "The Optimal Depletion of Exhaustible Resources", Review of Economic Studies, Symposium on the Economics of Exhaustible Resources, 1974.

11. Easterlin, R.A., "Population", in Contemporary Economic Issues, edited by N.W. Chamberlain, R.D. Irwin Inc., Homewood, Illinois.

12. Gale, D., "On Optimal Development in a Multi-Sector Economy", Review of Economic Studies, Vol. 34, 1967.

13. Gottlieb, M., "The Theory of Optimum.Population for a Closed Economy", Journal of Political Economy, 1945.

14. Harsanyi, J. "Cardinal Welfare, Individualistic Ethics and Interpersonal Comparison of utility", Journal of Political Economy, 1955.

15. Hauser, P.M., "World Population Growth", in The Population Dilemma, 2nd edition, edited by P.M. Hauser, Englewood Cliffs, N.J.: Prentice-Hall Inc.

16. Heer, D.M., "Economic Development and Fertility", Demography, Vol. 3, No. 2, 1966.

17. ________ , "Economic Development and the Fertility Transition", Daedalus, Vol. 97, No. 2, Spring 1968.

18. Ingham, A. and Simmons, P., "Natural Resources and Growing Population", Review of Economic Studies, Vol. XLII(2), No. 130, 1975.

19. Kingsley Davis, Human Society, New York, 1949.

20. Koopmans, T.C., "Proof for a Case where Discounting Advances the Doomsday", Review of Economic Studies, Symposium on the Economics of Exhaustible Resources, 1974.

21. ____________, "Some observations on 'optimal' economic growth and exhaustible resources", Cowles Foundation Discussion Paper No. 356, 1973.

22. Lane, J.S., "A Synthesis of the Ramsey Meade Problems when Population Change is Endogenous", Review of Economic Studies, Vol. XLII(I), January 1975.

23. _________, "Optimal Savings Policy when Labour Grows Endogenously: A Critique", Econometrica, Vol. 43, No. 5-6, September-November 1975

24. __________, "An Optimal Joint Population Savings Policy", Social Systems Research Institute Workshop Series, 1974, University of Wisconsin, Madison.

25. _________, "Optimal Economic Growth and Endogenous Population Change", Ph.D. Dissertation, Stanford University, 1972.

26. Meade, J.E., "The Growing Economy", Vol. 2 of The Principles of Political Economy, G. Allen and Unwin Ltd., 1968.

27. _________, "Trade and Welfare", Vol. 2 of The Theory of International Economic Policy, Oxford University Press, 1966.

28. Mirlees, J.A., "Taxation of Family Size". Journal of Public Economics, 1972.

29. Morishima, M., Theory of Economic Growth, Oxford University Press, 1969.

30. Narveson, J., "Utilitarianism and New Generations", Mind, 1967.

31. Pitchford, J.D., "Population and Optimal Growth", Econometrica, Vol. 40, No. 1, January 1972.

32. Pontryagin, L.S., Boltyanski, V.G., Gamkrelidze, R.V. and Mischchenko, E.F., The Mathematical Theory of Optimal Processes, New York, John Wiley and Sons, 1962.

33. Ramsey, F.P., "A Mathematical Theory of Savings", Economic Journal, 1928.

34. Rawls, J., A Theory of Justice, Clarendon Press (Oxford), 1972.

35. Ruff, L.E., "Optimal Growth and Technological Progress in a Cournot Economy", Technical Report No. 11, 1968, Institute for Mathematical Studies in the Social Sciences, Stanford University.

36. Sato, R., Davis, E.G., "Optimal Savings Policy when Labour Grows Endogenously", Econometrica, Vol. 39, No. 6, November 1971.

37. Sauvy, A., General Theory of Population, Weidenfeld and Nicolson (London), 1969.

38. Sidgwick, H., The Methods of Ethics, MacMillan (London), 7th edition, 1907.

39. Spengler, J.J., "Values and Fertility Analysis", Demography, Vol. 3, No. 1, 1966.

40. __________ , "Population Problems, Kyklos, 1950.

41. Uzawa, H., "Time Preference, the Consumption Function and Optimum Asset Holdings", in Value, Capital and Growth, edited by J.N. Wolfe. Edinburgh, Scotland: Edinburgh University Press, 1968.

42. Vickrey, W., "Utility, Strategy and Social Decision Rules", Quarterly Journal of Economics, 1960.

43. Vousden, N., "Basic Theoretical Issues of Resource Depletion", Journal of Economic Theory, Vol. 6, No. 2, 1973.

44. Wolfe, A.B., "On the Criterion of Optimum Population", American Journal of Sociology, 1934.

45. __________, "The Theory of Optimum Population", American Academy of Political and Social Sciences, 1936.

Vol. 59: J. A. Hanson, Growth in Open Economies. V, 128 pages. 1971.

Vol. 60: H. Hauptmann, Schätz- und Kontrolltheorie in stetigen dynamischen Wirtschaftsmodellen. V, 104 Seiten. 1971.

Vol. 61: K. H. F. Meyer, Wartesysteme mit variabler Bearbeitungsrate. VII, 314 Seiten. 1971.

Vol. 62: W. Krelle u. G. Gabisch unter Mitarbeit von J. Burgermeister, Wachstumstheorie. VII, 223 Seiten. 1972.

Vol. 63: J. Kohlas, Monte Carlo Simulation im Operations Research. VI, 162 Seiten. 1972.

Vol. 64: P. Gessner u. K. Spremann, Optimierung in Funktionenräumen. IV, 120 Seiten. 1972.

Vol. 65: W. Everling, Exercises in Computer Systems Analysis. VIII, 184 pages. 1972.

Vol. 66: F. Bauer, P. Garabedian and D. Korn, Supercritical Wing Sections. V, 211 pages. 1972.

Vol. 67: I. V. Girsanov, Lectures on Mathematical Theory of Extremum Problems. V, 136 pages. 1972.

Vol. 68: J. Loeckx, Computability and Decidability. An Introduction for Students of Computer Science. VI, 76 pages. 1972.

Vol. 69: S. Ashour, Sequencing Theory. V, 133 pages. 1972.

Vol. 70: J. P. Brown, The Economic Effects of Floods. Investigations of a Stochastic Model of Rational Investment. Behavior in the Face of Floods. V, 87 pages. 1972.

Vol. 71: R. Henn und O. Opitz, Konsum- und Produktionstheorie II. V, 134 Seiten. 1972.

Vol. 72: T. P. Bagchi and J. G. C. Templeton, Numerical Methods in Markov Chains and Bulk Queues. XI, 89 pages. 1972.

Vol. 73: H. Kiendl, Suboptimale Regler mit abschnittweise linearer Struktur. VI, 146 Seiten. 1972.

Vol. 74: F. Pokropp, Aggregation von Produktionsfunktionen. VI, 107 Seiten. 1972.

Vol. 75: GI-Gesellschaft für Informatik e.V. Bericht Nr. 3. 1. Fachtagung über Programmiersprachen · München, 9.–11. März 1971. Herausgegeben im Auftrag der Gesellschaft für Informatik von H. Langmaack und M. Paul. VII, 280 Seiten. 1972.

Vol. 76: G. Fandel, Optimale Entscheidung bei mehrfacher Zielsetzung. II, 121 Seiten. 1972.

Vol. 77: A. Auslender, Problèmes de Minimax via l'Analyse Convexe et les Inégalités Variationelles: Théorie et Algorithmes. VII, 132 pages. 1972.

Vol. 78: GI-Gesellschaft für Informatik e.V. 2. Jahrestagung, Karlsruhe, 2.–4. Oktober 1972. Herausgegeben im Auftrag der Gesellschaft für Informatik von P. Deussen. XI, 576 Seiten. 1973.

Vol. 79: A. Berman, Cones, Matrices and Mathematical Programming. V, 96 pages. 1973.

Vol. 80: International Seminar on Trends in Mathematical Modelling, Venice, 13–18 December 1971. Edited by N. Hawkes. VI, 288 pages. 1973.

Vol. 81: Advanced Course on Software Engineering. Edited by F. L. Bauer. XII, 545 pages. 1973.

Vol. 82: R. Saeks, Resolution Space, Operators and Systems. X, 267 pages. 1973.

Vol. 83: NTG/GI-Gesellschaft für Informatik, Nachrichtentechnische Gesellschaft. Fachtagung „Cognitive Verfahren und Systeme", Hamburg, 11.–13. April 1973. Herausgegeben im Auftrag der NTG/GI von Th. Einsele, W. Giloi und H.-H. Nagel. VIII, 373 Seiten. 1973.

Vol. 84: A. V. Balakrishnan, Stochastic Differential Systems I. Filtering and Control. A Function Space Approach. V, 252 pages. 1973.

Vol. 85: T. Page, Economics of Involuntary Transfers: A Unified Approach to Pollution and Congestion Externalities. XI, 159 pages. 1973.

Vol. 86: Symposium on the Theory of Scheduling and its Applications. Edited by S. E. Elmaghraby. VIII, 437 pages. 1973.

Vol. 87: G. F. Newell, Approximate Stochastic Behavior of n-Server Service Systems with Large n. VII, 118 pages. 1973.

Vol. 88: H. Steckhan, Güterströme in Netzen. VII, 134 Seiten. 1973.

Vol. 89: J. P. Wallace and A. Sherret, Estimation of Product Attributes and Their Importances. V, 94 pages. 1973.

Vol. 90: J.-F. Richard, Posterior and Predictive Densities for Simultaneous Equation Models. VI, 226 pages. 1973.

Vol. 91: Th. Marschak and R. Selten, General Equilibrium with Price-Making Firms. XI, 246 pages. 1974.

Vol. 92: E. Dierker, Topological Methods in Walrasian Economics. IV, 130 pages. 1974.

Vol. 93: 4th IFAC/IFIP International Conference on Digital Computer Applications to Process Control, Part I. Zürich/Switzerland, March 19–22, 1974. Edited by M. Mansour and W. Schaufelberger. XVIII, 544 pages. 1974.

Vol. 94: 4th IFAC/IFIP International Conference on Digital Computer Applications to Process Control, Part II. Zürich/Switzerland, March 19–22, 1974. Edited by M. Mansour and W. Schaufelberger. XVIII, 546 pages. 1974.

Vol. 95: M. Zeleny, Linear Multiobjective Programming. X, 220 pages. 1974.

Vol. 96: O. Moeschlin, Zur Theorie von Neumannscher Wachstumsmodelle. XI, 115 Seiten. 1974.

Vol. 97: G. Schmidt, Über die Stabilität des einfachen Bedienungskanals. VII, 147 Seiten. 1974.

Vol. 98: Mathematical Methods in Queueing Theory. Proceedings 1973. Edited by A. B. Clarke. VII, 374 pages. 1974.

Vol. 99: Production Theory. Edited by W. Eichhorn, R. Henn, O. Opitz, and R. W. Shephard. VIII, 386 pages. 1974.

Vol. 100: B. S. Duran and P. L. Odell, Cluster Analysis. A Survey. VI, 137 pages. 1974.

Vol. 101: W. M. Wonham, Linear Multivariable Control. A Geometric Approach. X, 344 pages. 1974.

Vol. 102: Analyse Convexe et Ses Applications. Comptes Rendus, Janvier 1974. Edited by J.-P. Aubin. IV, 244 pages. 1974.

Vol. 103: D. E. Boyce, A. Farhi, R. Weischedel, Optimal Subset Selection. Multiple Regression, Interdependence and Optimal Network Algorithms. XIII, 187 pages. 1974.

Vol. 104: S. Fujino, A Neo-Keynesian Theory of Inflation and Economic Growth. V, 96 pages. 1974.

Vol. 105: Optimal Control Theory and its Applications. Part I. Proceedings 1973. Edited by B. J. Kirby. VI, 425 pages. 1974.

Vol. 106: Optimal Control Theory and its Applications. Part II. Proceedings 1973. Edited by B. J. Kirby. VI, 403 pages. 1974.

Vol. 107: Control Theory, Numerical Methods and Computer Systems Modeling. International Symposium, Rocquencourt, June 17–21, 1974. Edited by A. Bensoussan and J. L. Lions. VIII, 757 pages. 1975.

Vol. 108: F. Bauer et al., Supercritical Wing Sections II. A Handbook. V, 296 pages. 1975.

Vol. 109: R. von Randow, Introduction to the Theory of Matroids. IX, 102 pages. 1975.

Vol. 110: C. Striebel, Optimal Control of Discrete Time Stochastic Systems. III. 208 pages. 1975.

Vol. 111: Variable Structure Systems with Application to Economics and Biology. Proceedings 1974. Edited by A. Ruberti and R. R. Mohler. VI, 321 pages. 1975.

Vol. 112: J. Wilhlem, Objectives and Multi-Objective Decision Making Under Uncertainty. IV, 111 pages. 1975.

Vol. 113: G. A. Aschinger, Stabilitätsaussagen über Klassen von Matrizen mit verschwindenden Zeilensummen. V, 102 Seiten. 1975.

Vol. 114: G. Uebe, Produktionstheorie. XVII, 301 Seiten. 1976.

Vol. 115: Anderson et al., Foundations of System Theory: Finitary and Infinitary Conditions. VII, 93 pages. 1976

Vol. 116: K. Miyazawa, Input-Output Analysis and the Structure of Income Distribution. IX, 135 pages. 1976.

Vol. 117: Optimization and Operations Research. Proceedings 1975. Edited by W. Oettli and K. Ritter. IV, 316 pages. 1976.

Vol. 118: Traffic Equilibrium Methods, Proceedings 1974. Edited by M. A. Florian. XXIII, 432 pages. 1976.

Vol. 119: Inflation in Small Countries. Proceedings 1974. Edited by H. Frisch. VI, 356 pages. 1976.

Vol. 120: G. Hasenkamp, Specification and Estimation of Multiple-Output Production Functions. VII, 151 pages. 1976.

Vol. 121: J. W. Cohen, On Regenerative Processes in Queueing Theory. IX, 93 pages. 1976.

Vol. 122: M. S. Bazaraa, and C. M. Shetty,Foundations of Optimization VI. 193 pages. 1976

Vol. 123: Multiple Criteria Decision Making. Kyoto 1975. Edited by M. Zeleny. XXVII, 345 pages. 1976.

Vol. 124: M. J. Todd. The Computation of Fixed Points and Applications. VII, 129 pages. 1976.

Vol. 125: Karl C. Mosler. Optimale Transportnetze. Zur Bestimmung ihres kostengünstigsten Standorts bei gegebener Nachfrage. VI, 142 Seiten. 1976.

Vol. 126: Energy, Regional Science and Public Policy. Energy and Environment I. Proceedings 1975. Edited by M. Chatterji and P. Van Rompuy. VIII, 316 pages. 1976.

Vol. 127: Environment, Regional Science and Interregional Modeling. Energy and Environment II. Proceedings 1975. Edited by M. Chatterji and P. Van Rompuy. IX, 211 pages. 1976.

Vol. 128: Integer Programming and Related Areas. A Classified Bibliography. Edited by C. Kastning. XII, 495 pages. 1976.

Vol. 129: H.-J. Lüthi, Komplementaritäts- und Fixpunktalgorithmen in der mathematischen Programmierung. Spieltheorie und Ökonomie. VII, 145 Seiten. 1976.

Vol. 130: Multiple Criteria Decision Making, Jouy-en-Josas, France. Proceedings 1975. Edited by H. Thiriez and S. Zionts. VI, 409 pages. 1976.

Vol. 131: Mathematical Systems Theory. Proceedings 1975. Edited by G. Marchesini and S. K. Mitter. X, 408 pages. 1976.

Vol. 132: U. H. Funke, Mathematical Models in Marketing. A Collection of Abstracts. XX, 514 pages. 1976.

Vol. 133: Warsaw Fall Seminars in Mathematical Economics 1975. Edited by M. W. Loś, J. Loś, and A. Wieczorek. V. 159 pages. 1976.

Vol. 134: Computing Methods in Applied Sciences and Engineering. Proceedings 1975. VIII, 390 pages. 1976.

Vol. 135: H. Haga, A Disequilibrium – Equilibrium Model with Money and Bonds. A Keynesian – Walrasian Synthesis. VI, 119 pages. 1976.

Vol. 136: E. Kofler und G. Menges, Entscheidungen bei unvollständiger Information. XII, 357 Seiten. 1976.

Vol. 137: R. Wets, Grundlagen Konvexer Optimierung. VI, 146 Seiten. 1976.

Vol. 138: K. Okuguchi, Expectations and Stability in Oligopoly Models. VI, 103 pages. 1976.

Vol. 139: Production Theory and Its Applications. Proceedings. Edited by H. Albach and G. Bergendahl. VIII, 193 pages. 1977.

Vol. 140: W. Eichhorn and J. Voeller, Theory of the Price Index. Fisher's Test Approach and Generalizations. VII, 95 pages. 1976.

Vol. 141: Mathematical Economics and Game Theory. Essays in Honor of Oskar Morgenstern. Edited by R. Henn and O. Moeschlin. XIV, 703 pages. 1977.

Vol. 142: J. S. Lane, On Optimal Population Paths. V, 123 pages. 1977.